eco barons

ALSO BY EDWARD HUMES

Monkey Girl

Over Here

School of Dreams

Baby ER

Mean Justice

No Matter How Loud I Shout

Mississippi Mud

Murderer with a Badge

Buried Secrets

eco barons

The Dreamers, Schemers, and Millionaires
Who Are Saving Our Planet

EDWARD HUMES

ecco
An Imprint of HarperCollinsPublishers

ECO BARONS. Copyright © 2009 by Edward Humes. All rights reserved.
Printed in the United States of America. No part of this book may be
used or reproduced in any manner whatsoever without written permission except
in the case of brief quotations embodied in critical articles and reviews.
For information, address HarperCollins Publishers,
10 East 53rd Street, New York, NY 10022.

HarperCollins books may be purchased for educational, business, or sales
promotional use. For information, please write: Special Markets Department,
HarperCollins Publishers, 10 East 53rd Street, New York, NY 10022.

FIRST EDITION

Designed by Jennifer Ann Daddio/Bookmark Design & Media Inc.
Title page image © Anne Kitzman/Dreamstime.com

Library of Congress Cataloging-in-Publication Data is available upon request.

ISBN: 978-0-06-135029-0

09 10 11 12 13   OV/RRD   10 9 8 7 6 5 4 3 2 1

To Donna

A man, after he has brushed off the dust and chips of his life, will have left only the hard, clean questions: Was it good or was it evil? Have I done well—or ill?

—JOHN STEINBECK, *EAST OF EDEN*

CONTENTS

PART THREE

waiting for thoreau

PART FOUR

lone wolves

eco barons

the plan—
buy low, sell never

The little Cessna darted out from under the clouds and Doug Tompkins got his first glimpse of Reñihue Fjord, a misty blue gem in southern Chile, where the longest country in the world is pencil thin and the snow-capped Andes tumble toward the sea like an army finishing a long march. A colony of seals basking on the fjord's rocky beach lifted their heads skyward to bellow at the passing plane as Tompkins banked inland, away from the sparse ranch that occupied the brown and narrow coastal flats. Now he was over a majestic river valley, woods seemingly untouched by saw, ax, or torch, and though he was not a religious man, it was impossible not to think of fabled Eden in this damp and glorious setting, the world's last intact temperate rain forest. Everywhere he looked, a raw, primordial wilderness filled his senses. Wet, sun-dappled, mysterious, it was colored in shades of green and blue not even Rousseau imagined for his most vivid jungle canvases—colors lit from within and pulsing with life so deep they seemed of another world.

Tompkins's world was a very different one, the world of industry and commerce, of image over substance—the fashion business. Between 1970 and 1989, he had built a catalog of knockoff clothes into a distinctive brand, and then had made the brand into a symbol of what people yearned to be, and then he had become an icon himself, a personification of his company, the hottest label of the era: Esprit. That made him a visionary, or so people said, a perception he encouraged by titling himself "image director" instead of CEO. But now it was 1990 and his company was in turmoil, his marriage had been destroyed, and the pride he had always taken in his craft had long since ebbed. Cruising now above a turquoise river twisting through virgin forest, Tompkins found himself wondering, not for the first time, just what the hell he had been thinking all those years he spent deciding what color socks teenage girls would wear, or how tight or low-cut their jeans would be next fall. Accolades and success in the world of fashion didn't seem to mean much here in the heart of Patagonia, the lush southern third of the South American continent, named by Magellan for the Patagons, mythical giants he swore lived there. Patagonia remains one of the last completely wild *big* places left on earth, home to sleek, elusive pumas and tiny *pudu*—rare deer scarcely bigger than poodles—and 3,600-year-old *alerce* trees, towering cousins to California's sequoias. The tallest *alerces* still standing are among the oldest organisms on earth. They bore witness to the rise of civilization. And like every other wild thing in Patagonia, they are in danger.

Tompkins had been an outdoorsman all his life: a daring whitewater kayaker; a skier with aspirations to compete in the Olympics; a serious mountain climber who once spent four weeks holed up in an ice cave with four buddies, waiting out an epic storm until they could finally blaze a new trail to the summit. That had been in Patagonia, too, twenty-two years and a lifetime ago, and ever since, he had always built into his business model three, four, or five months in the wild somewhere. MBA, he called it: managing by absence. He had done it when Esprit was little more than a young married couple in San Francisco selling flowery hippie dresses out of their station wagon in 1970, and he had done it as CEO of a billion-dollar global fashion empire. More often than not, with the whisper of an ancient forest and the pelting drumbeat of rain so frequent it has to be measured in meters, the wild place that called to him was Patagonia. He understood what Pablo Neruda, Chile's Nobel laureate poet, meant in writing, "Anyone who hasn't been in the Chilean forest doesn't know this planet."

For years, Tompkins had tried to nurture his inner environmentalist through his business. He regularly wrote checks to respectable conservation groups. He built an urban park near Esprit's headquarters in San Francisco, a splash of green and transplanted redwoods beloved by the neighborhood's kids and dog owners. There were the essays he slipped into the company catalogs, urging people to embrace healthier lifestyles, recycle more, consume less. And when his old mountain-climbing buddy, the founder of the clothing company Patagonia, called him in 1988 and asked him to chip in to buy and preserve some fast-vanishing Chilean forest land, he impulsively agreed. Tompkins's $50,000 helped bankroll a nature preserve filled with endangered araucaria trees,[1] towering evergreens with a whimsical Dr. Seuss look, commonly called the monkey puzzle tree. Just like that, a park was born instead of a field of stumps. No red tape. No bureaucrats weighing commercial interests against the environment—the sort of lobbyist-driven balancing act that had turned whole swaths of the United States' protected national forests into lucrative fiefdoms for timber, mining, and oil concerns. Six thousand miles south, however, it seemed all you had to do was write a check, and it was done. No one would ever cut down those araucaria, living fossils that have survived as a recognizable species since the age of the dinosaurs, impervious to all threats but one: the age of man.

Suddenly it seemed so clear to Tompkins. He had been on the wrong side too long, running a global fashion business with factories in Hong Kong, boutiques and superstores worldwide, and offices in a dozen countries. He shipped clothes all over the earth, using artful images and beautiful models to persuade people to consume things they didn't need, then eagerly replace them each season with new things before the old were even close to worn out. He had tried to compensate with his donations and his essays, but Esprit was never going to be green or sustainable or good for the earth, and neither was Tompkins—so long as he was part of it. And so, amid crises and take-over bids at Esprit, as the vulture investors circled and his empire of image began to collapse, he rounded up a few close friends and, with no explanation beyond his need to get away for an adventure, they flew to Patagonia in a pair of small planes. Tompkins's company and his marriage were besieged and yet, as the Chilean rain forests slid by beneath his plane, his friends could see that he looked happier than they'd seen him in years.

Before leaving San Francisco, he had called the Chile-based activist who brokered the deal to save the araucaria trees, and Tompkins gleaned the piece of information that would change his life: A lot more of Patagonia could still be bought cheap. Old-growth forests could be had for as little as twelve dollars an acre, and one such area was Reñihue. When Tompkins completed his flyover and started talking to a real estate agent, he found that a broken-down ranch next to the fjord, along with all its spectacular surroundings, was for sale. The seller would even throw in the cattle. He could have 24,700 acres for $600,000.

Tompkins tried to keep a poker face: $600,000? You couldn't buy a condo in San Francisco for that kind of money. In Chile—remote, beautiful, wild Patagonian Chile—$600,000 would buy him nearly forty square miles. Hell, the whole city of San Francisco was only forty-seven square miles. In what kind of crazy upside-down world was a tiny three-bedroom piece of the paved-over, used-up urban landscape of San Francisco worth more than the most beautiful wilderness on earth?

"What about the volcano?" Tompkins asked. In the distance, visible from the ranch house, stood the cone of the dormant 8,000-foot volcano Michinmáhuida, snowcapped and heavily forested.

"That's included, too, señor." And there were many more areas on the market right next door and up and down the coast, the agent added. The deposed Chilean dictator Augusto Pinochet's cronies had bought up the land years ago, trusting the old tyrant to develop Patagonia with roads and factories and thereby enrich his friends. Now Pinochet was out and facing indictment, and land newly deemed worthless was priced to move. "It's all for sale."

His own volcano—it was almost too much to grasp. You couldn't preserve even the tiniest parcel for that kind of money in the United States, but here the possibilities were endless. Tompkins was in his fifties, his hair gone silver, his face weathered, but he still had time. If he sold his interest in Esprit, the millions he'd earn could have a huge impact. He could buy more land than he had ever imagined, with plenty left over to stir up some trouble at home, too. He might finally endow that foundation he had long coveted. He could publish books, run full-page ads in *The New York Times*, dole out grants to environmental groups. And in Chile, he could buy paradise. He could save paradise. He could *live* in paradise.

Here, near the bottom of the world, the fashion mogul saw a microcosm for all of humanity's environmental ills—and a laboratory for finding ways to fix them. The big threats were all here: destruction of forests and habitats, dying rivers and ocean coasts, topsoil erosion, brutal extraction of resources and energy without regard for life or landscape. There were displaced populations, lost cultures, and mass extinctions already under way—not just of rare animals but of critical species that pollinate crops, control pests, and clean the air—a rate of dying out that hasn't been seen in 65 million years, when the once-dominant dinosaurs perished. Underlying it all, there was the inexorable creep of climate change and an utter unwillingness of leaders and populations to confront it in a meaningful, honest way. In Patagonia, as in the United States and throughout the world, nature was under siege as never before, though the difference was that most of Patagonia could still be saved. For Tompkins the issue was no longer about saving a snail darter or a spotted owl; it was about preserving a viable world for his children and grandchildren. It was about averting a disaster of man's own making—one he had been heedlessly helping along for twenty years. His CEO's outsize ego would be channeled in a new direction: doing penance.

He would be making the ultimate flip-flop, from global capitalist to anticonsumerism ecologist. Inevitably there would be opposition once his intentions became clear—he could foresee being called an arrogant gringo, a colonizer, an environmental extremist, a hypocrite who had made his own millions but wanted Chileans to forgo progress and remain peasants. This was an age-old response. Buying land to exploit it, pave it, mine it, or build on it was welcome everywhere from Jackson Hole, Wyoming, to Tierra del Fuego at the tip of South America, but preserving it . . . that was suspect and subversive. Sixty years earlier in Wyoming, mobs had burned John D. Rockefeller Jr. in effigy when he proposed protecting the land that eventually became Grand Teton National Park. Now they would go to war against anyone who tried to take it away. At Esprit, Tompkins knew, his colleagues and employees would think he had lost his mind. But he thought he had finally found himself.

He made the deal that same day. Reñihue—pronounced, coincidentally and appropriately, "rainy way"—was his. And soon Esprit was a memory and he was living half the year on the fjord, holed up in a tiny cabin with no electricity or phone, working to restore the ranch, working with his hands, pulling stumps, preaching the virtues of a local economy in lieu of a global economy. And he acquired more land, a great deal more land. There, in the first piece of an empire-to-be, he began a new life of conservation and activism that would make him one of the most influential eco barons of his time—perhaps the most influential of them all.

There is a plan. It is audacious. It is huge. It has provoked a silent, high-stakes war that most Americans know nothing about, though it will affect every one of us and generations to come.

It is already unfolding, a secret plan to save the earth.

In an era in which government has been either broke, indifferent, or actively hostile to environmental causes, a band of visionaries—inventors, philanthropists, philosophers, grassroots activists, lawyers, and gadflies—are using their wealth, their energy, their celebrity, and their knowledge of the law and science to persuade, and sometimes force, the United States and the world to take a new direction. They use lawsuits, charitable foundations, land trusts, mass protests, armies of schoolchildren, and billions in corporate profits. They use pictures of beached whales, videos of starving polar bears, and murals of vast fields of stumps where towering pine forests once stood. They use anger and fear, and they use hope.

They are following in the tradition of Rockefeller, Carnegie, and the storied initiatives that saved Yellowstone, Yosemite and the Grand Tetons from the bulldozer and the Grand Canyon from the dams. Their goals are to save wilderness and rain forests from destruction, to slow the rising tide of mass extinctions, and to "re-wild" large swaths of land and migratory corridors where development has long intruded. They are pushing new (and old) green technologies, farming methods, and sources of energy, beginning the process of weaning the country and the world from the fossil fuels that drive global warming, whether government leaders like it or not. They seek to show, in deeds and words, that it is possible to strike a better balance between consumption and conservation and still prosper—to save the world, piece by piece, species by species, place by place.

They are the eco barons.

They are the antithesis of the robber barons of the "gilded age," those clever and powerful railroad, land, and banking barons who used any means necessary to accumulate wealth and power—transforming an America of individual farmers and merchants into a global industrial powerhouse of prosperity. But along the way, they also laid waste to the rule of law, human rights, natural resources, and America's vast unspoiled rivers, forests, and landscapes. Today's eco barons, like the robber barons before them, are also accumulating vast amounts of land, building new industries while transforming old ones, pioneering new technologies, and taking advantage of legal loopholes and underused statutes. But instead of making money, these latter-day barons have vowed to save the natural environment their predecessors despoiled, and in the process to secure a better future for humankind. Some are fabulously wealthy, some live off modest paychecks even as they realize outsize ambitions—it is their impact, not their bank accounts, that makes them eco barons.

Among them are Kieran Suckling and Peter Galvin, the two former forest service employees who launched America's most effective environmental law firm, consistently outwitting the best private, corporate and government attorneys in the world to protect millions of acres of forest and wilderness. The eco barons' public profiles range from famed media figure Ted Turner, who owns—and has preserved—more land in the United States than any other individual American, to obscure inventor Andy Frank, a university professor who is giving the world a better, greener, cheaper way to drive. There is Terry Tamminen, once a pool cleaner, then head of California's Environmental Protection Agency, now a sustainability consultant, who has somehow persuaded two Republican governors to create the most environmentally friendly states in the union. There is Carole Allen, a single mom and volunteer who marshaled an army of schoolchildren to shame fishing fleets into saving endangered species. And there is Roxanne Quimby, the founder of a cosmetics powerhouse, Burt's Bees, who is fighting to preserve the great Maine Woods—an inspiration for Henry David Thoreau 150 years ago and now one of the largest forests left standing in the United States.

Then there are Doug Tompkins and his wife, Kris, the Rockefellers of their generation, the eco barons whose impact seems to reach farthest. Their network, influence, and support for green causes stretch throughout the nation and the world, sometimes highly visible, often out of sight yet exerting a profound impact on the environment and the daily lives of Americans. They try to lead by example, running ranches, resorts, private parks, organic farms, industries, and entire communities in a renewable, energy-efficient, nonpolluting, wildlife-friendly, and *profitable* fashion. They say America needs to follow this example in a big way, to create the conservation and energy equivalent of President Kennedy's declaration in 1961 that the nation would reach the moon within the decade. Action for conservation and against climate change need not take a backseat to economics and national security, they argue, rejecting the very notion that there is a conflict between the two. The Tompkinses and their fellow eco barons not only say that immediate action to avert disaster is vital to security and prosperity—there are many voices in that chorus, including the new president's—but are actually taking action, and asking the rest of the world to catch up.

And so the eco barons' plans stretch from the majestic Chilean fjords to the Alaskan wilderness preserves, with their coveted oil reserves deep below virginal landscapes occupied by moose, polar bear, and seal. They span North America from the Everglades to the Adirondacks to Newfoundland, swamp to mountaintop to forest, and stretch westward through the plains states, the Big Sky country, and the deserts and forests of the Southwest. Those plans reach from the jade shadows of Patagonia's vast temperate rain forest—the last one of its kind in the world, a treasure trove for conservationists and logging companies alike—to the granite peaks and riotous wildflower fields of the Tejon Ranch just north of Los Angeles, one of California's last wild places, a magnet for biodiversity and endangered species. Developers want to build a new city out of nothing there, 75,000 people (plus malls, gas stations, and supermarkets) plunked down instantly in a pristine wilderness where some of the last California condors forage near cliffs they have occupied for thousands of years.

For decades it has been possible to say, as each new suburb, freeway, and industrial park swallowed up yet another million acres of wilderness, that there was always plenty more where that came from. Now the wild areas are running out, and irreplaceable habitats are under threat. Vital natural partners in the human ecosystem—bees and birds and old-growth trees—are being wiped out with the same certainty that there will always be more, but this certainty is misplaced. The rate of extinctions in some places has exploded. Instead of the average extinction rate—one species a year going extinct per million species alive (a rate that had persisted for many thousands of years as a normal aspect of evolution and natural selection)—some regions are seeing 1,000 times that rate of extinctions in a year. In the South American rain forests, the rate has reached 10,000 extinctions per million species a year. Entire lines of evolution, of DNA, of potential cures for diseases and food for the masses are gone. Many of the creatures that pollinate the crops we depend on are vanishing. One in eight bird species faces extinction—one in eight, gone. Turtles and other amphibians? One species in three is at risk. Three-quarters of flowering plant species, critical parts of the food chain, are also at risk. The oceans are dying, and with them the food supply for billions of people: 90 percent of the big fish—cod, tuna, and swordfish—that once swam the ocean depths are gone. This is not a natural process. It is a human one.[2]

There are those who wish to continue business as usual, who say that changing the way we drive, build, shop, make electricity, grow food, and cut down our forests would be too inconvenient, too disruptive, too costly. No one knows for sure what the future holds, these voices argue, so why disrupt everything when we can continue burning coal (and removing whole mountaintops to get it), when we can

drill offshore for oil and get the same twenty-five miles a gallon that
we got when Ronald Reagan occupied the White House and the most
popular car in America was an Oldsmobile? It is tempting, especially
during times of economic turmoil, to embrace such wishful thinking, to avoid the hard choices. For the first eight critical years of the
twenty-first century, America was ruled by a president who thought
this way, who blocked attempts to save wilderness, to protect endangered species, to reduce the greenhouse gas emissions that cause global
warming, and to curb the country's insatiable thirst for oil. Instead,
the president and his allies in Congress took the opposite tack: drilling for oil in pristine wilderness areas, logging in national forests, re-
fusing to obey environmental laws, even giving enormous tax credits
to subsidize the purchase of one of the least efficient, most wasteful
passenger cars ever built, the Hummer. Such a tax policy provides an
unintentionally apt metaphor for America as a whole, the most wasteful nation in human history, with 5 percent of the world's population
consuming nearly 25 percent of its energy and producing 25 percent
of the world's greenhouse gases. Americans love the idea of conservation, but few of us have vigorously practiced it since World War II,
when rationing was imposed. Our mass transit system is, except for
pockets in the Northeast, an international joke. Our cars, our appliances, and even our lightbulbs are less efficient than those of the rest
of the world. We actually waste more energy in our homes, cars, and
businesses than we import in the form of foreign oil, and this is one
reason why we consume nine times as much electricity per capita as
the world average. We offer tax breaks, investment incentives, and
direct subsidies for sprawl, pollution, and waste that far outstrip any
incentives to conserve, to build smart, and to pollute less—in effect,
our policies, from urban planning to homeland security, almost seem
designed to encourage inefficiency and environmental degradation.
Industrial agriculture receives government subsidies on a huge scale
that make it cost-effective for farmers to burn the equivalent of 400
gallons of oil a year for every person fed[3]—so that simply feeding

America expends 120 billion gallons of oil a year, more than most countries consume for all purposes. Overall, America consumes 1.3 trillion gallons of oil a year, about two-thirds of it for transportation. In order to sustain this level of consumption, we transfer our wealth on an epic scale to foreign oil suppliers, a pattern that remained unchanged even though the attacks of 9/11 were financed with some of those same oil dollars. Developing nations worldwide are busily cutting and plowing over rain forests, extracting resources, and killing off species in order to feed America's and their own growing appetite.

The new eco barons have a word for this: unsustainable.

This is another way of saying that, however uncomfortable facing up to it may be, business as usual cannot continue without incurring certain disaster. The eco barons offer a message of change—not in the form of rhetoric, but as a series of national and global projects that are, right now, saving parts of the planet. Acting individually and sometimes in concert, these remarkable visionaries are showing the world that nature can be nurtured; that forests can be sustainably logged; that energy efficiency and conservation can protect the environment and save enormous amounts of money; that alternative methods of farming, driving, and making electricity can enrich our lives and our bank accounts while reducing the threat of climate change. Many Americans were awakened to the severity of global warming by Al Gore's landmark documentary *An Inconvenient Truth*—and despaired. The eco barons are writing the next chapter in the story, and theirs is a message of hope: The world can be saved. Civilization, society, prosperity, a good life, a stable future—all are still there for the making. The real question is not if we can, but if we *will*: if the world will follow the lead of the eco barons, or if the naysayers and deniers will win, with business as usual continuing until disaster is inevitable. The world as we know it—the stable sea levels, temperate weather, and ample natural resources that allowed civilization to flourish and grow—will change in a few short years if we don't follow the eco barons' lead. Resources will be used up. Extinctions will go out of control. Ice packs will melt and coastal regions will flood. Droughts and extreme storms will plague once temperate regions. Fuel and food supplies will shrink. And life will become very, very hard for much of the world. Soon, the eco barons say, inaction may lock us into that fate, and it will be too late.

"Sometimes, it seems it already *is* too late," Doug Tompkins says. His grandest of plans, his hope to restore and reconnect wilderness areas around the world, stretches many years in the future, 100 years perhaps; it is a plan for his grandchildren and great-grandchildren to see through. He lives with the knowledge that the eco barons' battle to save the natural world will not be won in his lifetime, though certainly he could witness it being lost. "Sometimes we despair, we see nothing changes, and we think, yes, it probably is too late. But then we keep going, because, who knows, we may be wrong."

Part One

out of fashion

The members of this new ruling class were generally, and quite aptly, called "barons," "kings," "empire-builders," or even "emperors." They were aggressive men, as were the first feudal barons; sometimes they were lawless. . . . [They sought] to organize and exploit the resources of a nation upon a gigantic scale, to regiment its farmers and workers into harmonious corps of producers, and to do this only in the name of an uncontrolled appetite for private profit. Here surely is the great inherent contradiction whence so much disaster, outrage and misery has flowed.

—MATTHEW JOSEPHSON, *THE ROBBER BARONS* (1934)

"It is pretty hard for a country to turn down a gift of 300,000 hectares," explains Douglas Tompkins, 64, the de facto dean of this new class of eco barons, who has spent the past decade and $US 200 million spearheading a new movement called Wildlands Philanthropy.

—*SYDNEY MORNING HERALD,* 2007

reaching the summit

His friends say it makes perfect sense, this transition from the fashion world to saving the world. All the pieces were there for years, hiding in plain sight. Still, none of them—in some ways, not even Tompkins himself—saw it coming. The metamorphosis of the CEO of Esprit fits only in hindsight, as a journey that mirrors the changing priorities, assumptions, and points of view at the heart of many executives' and corporations' greener thinking in the twenty-first century—the principal difference being that Esprit's chief image maker got there twenty years ahead of the pack.

Douglas Tompkins grew up in the village of Millbrook, New York, a Hudson Valley enclave of tree-lined roads, rolling green pastures, and large homes with horse barns and plenty of land. His ancestors arrived on the *Mayflower*. In 1943, when Tompkins was born into a world at war, Millbrook was already known for its moneyed inhabitants, understated country elegance, and walled estates.

Today it is one of the wealthiest towns in New York state, and such diverse figures as Jimmy Cagney, Mary Tyler Moore, Katie Couric, and Timothy Leary (the apostle of LSD) have called it home.

Tompkins's mother was a decorator and his father was in the antique business—high-end, appointment-only antique dealing, which involved combing the region for museum-quality pieces and works of art in a private plane and seeing clients in their homes and galleries. If Doug Tompkins's flashes of warmth and gentleness, as well as his deep attraction to forests and nature, come from his soft-spoken mom, his most obvious trait—stern certitude—comes from his dad. A tough, demanding, tasteful man with a sharp eye for quality and style, the elder Tompkins expected no less from his sometimes unruly son. He presented young Doug, age ten, with a book that explained how to distinguish between good and bad specimens of Chippendale and Hepplewhite furniture—and he expected the boy to read and discuss it.

The son may have inherited the father's eye for design and style, but the antique dealer's traditionalist views and sense of order were another matter. The respected boarding school his parents chose for his high school years—Connecticut's Pomfret School, whose students would include another future eco baron, Robert F. Kennedy Jr.—could not contain Tompkins. The headmaster expelled him in his senior year for rule-breaking and rebelliousness when he failed to come back after a weekend—for the tenth time. "I wasn't great on heeding authority," Tompkins says now, shrugging at the memory. "I'm still not too good at that."[1]

At age seventeen, at the dawn of the 1960s, he gave up on high school, taking off for Colorado to ski bum, mountain climb, and go "adventuring," as he calls it. The outdoors mattered to him most: He had started rock climbing when he was twelve in the Shawangunk Mountains, a favorite New York spot for climbers seeking a challenge, and by fifteen he was skiing and climbing mountains during family trips to Wyoming. In Aspen, he waited on tables and worked in ski shops, taking two jobs at a time during the seasonal holiday crunch, squirreling away all his money, saving for his next journey, passing himself off as older, relishing being on his own. The tips were good, but even better were the free staff lodgings, meals, and ski passes, which meant his expenses hovered near zero and the slopes were wide open to him.

After a year spent in Colorado hoarding cash, he took off for Europe, where he first climbed the Alps. Then he traipsed through the Andes in South America, making his first visit to Patagonia. Even then, eighteen and heedless, he recognized the rain forests of Chile and Argentina as special places, and he was in no hurry to leave. He stretched his money by hitchhiking, camping, and eating next to nothing while roaming the landscape. When his money finally gave out, forcing him back to the states to find more work, he did not settle back in Aspen. It was 1962, John Kennedy was the president of an America not yet tainted by assassination or Vietnam or Kent State or Watts. Where else would a young man from back East with no ties and no plans beyond making a run at the U.S. Olympic ski team go but California? He put out his thumb and headed west.

He landed on the outskirts of Tahoe City, where he found plenty of work at the ski resorts during the snowy season as a trail and mountain guide, leading to his first business venture, the California Mountaineering Service. During the summer he worked in construction or as a tree topper, taking down the big Douglas firs that were interfering with power lines and summer homes. Because the trees were so large and so close to houses, chopping them at the bottom was out of the question; tree toppers had to scale the trunks with cleats and harnesses and remove the tree piece by piece, working from the top down—difficult, dangerous, and time-consuming work, paying good money for a high school dropout.

Hitchhiking home from a job one summer day, Tompkins watched as a vivacious twenty-year-old with a blond ponytail pulled over and opened the door of her Volkswagen. Susie Russell peered out at the wiry, dark-haired Tompkins, discerning a certain rugged charm about him, though in those days, she would have offered a ride to just about any hitchhiker—everybody did back then. She was working that summer as a keno girl at the Nevada Lodge, just over the state line, where she had used a phony ID to get around the casino age restriction of twenty-one so that she could run bets and winnings back and forth from the keno tables.

Tompkins climbed in and, while introducing himself, boasted that he was a Harvard man. Russell was attractive and smart and Tompkins wanted to impress her, but he miscalculated. She had much more in common with Tompkins's real résumé than with his imagined one. Her dad had been a well-known real estate developer and San Francisco's betting commissioner back in the day of legal gambling parlors. Her mother, an artist, had complained about her headstrong daughter's embrace of the counterculture and her lack of interest in college. Susie went to several top private and public high schools in San Francisco but she, like Tompkins, left high school without her diploma: Her principal at Lowell High School said she would not be welcome at graduation because she had been "too wild" at the prom. So the lie about Harvard didn't gain Tompkins any traction—quite the opposite. Who needs a pretentious lumberjack? Susie asked, suggesting he could hop out of the car then and there. But he shook his head, and she dropped him where he wanted to go.

"He seemed so arrogant!" Susie recalls more than forty years later. "Still, there was something about him. He was unique. You couldn't miss that. Maybe we were attracted to each other's uniqueness. Neither of us wanted ordinary lives."

In a bigger city, that ride probably would have been the end of the story. But Tahoe was small, and it turned out they moved in the same circles. They both skied, they both craved adventure, they both resisted conformity—and they kept bumping into each other in stores, at parties. One time he hit her up for a sixty-dollar loan, and she said

OK, scraping together her precious tip money and handing it over. He took this (correctly) to be a good sign.

Gradually they became better acquainted, two strong personalities who were not afraid to butt heads—each was the sort that enjoyed a bit of conflict, and even thrived on it. One day, without prelude, Tompkins suggested, "Let's go down to Mexico." This was what Tompkins has always been good at, his friends recall: He could make the sudden impulse to drop everything and embark on a new adventure sound not outlandish, but irresistible. This is why people were drawn to Tompkins—you never knew what might be in store, where he might suddenly disappear for months at a time, or whether you might disappear with him. Impulsively, Susie said yes, she would come along.

They spent weeks tooling around Mexico in a VW microbus, both up for anything, surprised at how easy it was to be together, until suddenly all of autumn had slipped by, she could no longer imagine being apart, and Doug proposed. They were just winging it, two kids feeding each other's souls, as Susie later described the time. She said yes.

They worked another season in Tahoe—ski patrol, waiting tables, doing pretty much anything to make money. By 1963, they were ready to head to San Francisco, the place Susie knew and loved best, their relationship already assuming what would become a permanent pattern: passionate conflict, passionate reconciliation, and long months apart as Doug continued adventuring and training for an Olympic tryout—marriage and kids (soon enough) notwithstanding. This was the deal going in: They would have full lives together, and full lives apart. Susie knew she was not acquiring the sort of husband who came home every day at five, nor did she want one who did.

Even before the move from Tahoe, Tompkins had fallen in with a crowd of hotshot climbers who were in the midst of creating what would later be referred to as the "golden age" of climbing in Yosemite

National Park—blazing new routes up the park's most difficult (and in many cases, unconquered) summits, hanging out beyond their two-week camping passes, then dodging park rangers who wanted them to clear out. Some of the world's best climbers congregated at Yosemite back then, claiming the park's Camp 4 as their domain and headquarters. They were nicknamed the "Valley Cong"—the guerrilla climbers. They saw themselves as creating a new aesthetic for climbing that emphasized style, and they took pride in the fact that there was absolutely no economic benefit to climbing. It was pure. It was outside society. It was an adrenaline-charged, risk-taking endeavor, different every time. And it was a unique way of looking at the world and at achievement.

Tompkins befriended and started climbing and surfing with one of the Valley Cong's leaders, Yvon Chouinard, a southern California rock rat whom Tompkins, then fifteen, had first met in 1958 when he was climbing in the Shawangunks. Chouinard, five years older, was making a name for himself by climbing a series of peaks in Yosemite without the aid of ropes and without the time-consuming "siege tactics" applied by earlier generations.

Despite their climbing skills, Chouinard's and Tompkins's first adventure together was almost their last: a climb in 1962 up the north face of Mount Temple in the Canadian Rockies, an 11,624-foot peak in Banff National Park known for its treachery. At that time, its north face had never been scaled. Halfway up, as they camped on an exposed snowfield, bad weather arrived. They were vulnerable to wind-induced rock and snow avalanches, and Tompkins suggested they get out, even if it meant a harrowing climb down in darkness. They made it off the mountain by dawn, just in time to watch a rock slide explode onto their abandoned campsite. They've been best friends ever since, more than forty years, adventuring together on every continent, including Antarctica.

Chouinard saw in Tompkins the kind of informed fearlessness that makes a great climber and adventurer—and, as it turns out, a successful businessman: a willingness to leap, married to a certainty that you would reach the other side. "Fearless" was a word Tompkins has long used to describe how he wants to live, an expression of his two long-standing obsessions: going "where the ordinary human doesn't go," and being "world-class" at whatever he attempts to do. It's the difference between simply breathing versus inhaling so deeply your lungs feel about to burst, he once observed. His wife Susie put it less metaphorically: "Doug is always trying to outperform himself."

Chouinard, in turn, was an inspiring figure to Tompkins: tough, good-humored, supremely competent, inventive, outspoken. Around the time of Tompkins's move to California, Chouinard was in the process of reinventing climbing gear with a secondhand forge, an old anvil, and metal he had salvaged from old harvester blades. It was a little business, Tompkins saw, but it was genius—because no one else had thought to do it.

Back then, climbers hammered pitons into rock walls for support as they moved upward. Most pitons were made of soft iron; they were used once and left behind. But climbing the sheer rock faces of Yosemite and similar vertical mountain faces called for hundreds of piton placements—one-use pitons were impractical. So Chouinard taught himself to be a blacksmith and made his own hard steel pitons out of old stovepipes and other junkyard scrap, which he tested out on some of the earliest ascents of Yosemite's Lost Arrow Chimney and Sentinel Rock. Pretty soon other climbers heard about the Valley Cong's mountaineering blacksmith and began demanding his pitons, which he sold for $1.50 apiece out of the back of his rattling old car.

Chouinard traveled the country in this way, climbing, surfing, making and selling his scrap-metal pitons, and working summers as a guide in Wyoming, where a young broadcast journalist and future network anchorman, Tom Brokaw, became his climbing student and friend. The rest of the year Chouinard survived solely on the meager income from his piton sales, running so low on funds at one point that he had to live off a very inexpensive case of dented cans of cat food

he bought at a surplus store, supplemented by the occasional squirrel. He'd haul his anvil out on the beach after a few hours in the water
and pound out more pitons, earning a few more bucks. He invented
another, extremely small piton he called the "realized ultimate reality
piton" (RURP), which soon became state of the art in rock climbing,
and his financial outlook began to look better. By 1965, there was
enough demand for Chouinard's climbing gear that he and a partner
replaced his anvil-in-a-car manufacturing setup with a blacksmith
shop in Ventura, California, in an old building with corrugated steel
walls: Chouinard Equipment. They offered a spectrum of redesigned
and improved tools for serious devotees of rock, mountain, and ice
climbing.

Around the same time Chouinard decided climbing gear needed
to be reinvented, Tompkins was feeling similarly inspired by a lack of
high-quality, lightweight, readily available camping gear for climbers and serious outdoorsmen—expedition-grade sleeping bags, tents,
packs, and other portable equipment. So the Tompkinses started
importing and selling the equipment, including Chouinard's gear,
mostly out of their car at climbers' camps and parks, cash and carry.
They supplemented their income with another business, Recreation
Unlimited, a kind of summer adventure camp for boys. Tompkins,
sometimes with Susie in tow and other times on his own, would haul
kids on climbing, hiking, and camping getaways throughout California and Mexico. They advertised in *Sunset* magazine, and in those in-
nocent days, unburdened by licenses, regulation, or massive liability
insurance policies, they would persuade parents to send kids off for
outdoor adventures with Doug the Mountain Man.

It was a modest business and their budget was always tight, but
then a new opportunity arose: a friend with a ski shop in the Bay
Area told them the business was failing and he was going to close
down. The Tompkinses offered to rent the space to set up a store for
their climbing and camping equipment. They borrowed $5,000 and
in 1964 started The North Face, a mail-order and retail company,

located in San Francisco's North Beach area, across the street from beat poet Lawrence Ferlinghetti's legendary bookstore, City Lights. Tompkins chose the name because the north face of any mountain above the equator is usually the coldest and most difficult to climb— as he and Chouinard had experienced firsthand on Mount Temple. The logo was a stylized drawing of Half Dome at Yosemite, a favorite of the Valley Cong.

Tompkins stocked high-end mountaineering and camping equipment he bought wholesale from Europe, as well as Chouinard's technical climbing gear, the two entrepreneurs combining forces and helping each other's business grow. At the time, North Beach was a mix of music, culture, and the occasional topless club, where Allen Ginsberg, Alan Watts, and Jefferson Airplane were neighborhood regulars and the Tompkinses bought their first house. The North Face store stood out in this setting; the aesthetic that would later shine through at Esprit got its start with the store Tompkins designed, dominated by large photo murals of mountains, forests, and snowfields, along with the clean, natural wood counters and floor. A huge Ansel Adams print, borrowed from a local printer where it had been rolled up and collecting dust, covered one wall. The other walls were paneled with gorgeous distressed redwood that Doug had retrieved by driving up to the chicken farms that once prevailed in the northern California town of Petaluma, where he volunteered to tear down residents' aged, disused chicken coops and haul away the wood. The effect was sophisticated and attractive; the sparse selection of merchandise, though high quality, was almost a sideshow. The emphasis was on where you would go and how you would feel with that equipment, what it was like to be in the mountains or hanging from some vertical slab of rock—everything was filtered through Doug Tompkins's sensibilities. He had an instinct for creating a compelling image, something that comes naturally, perhaps, to climbers, who tend to project a very definite image of themselves.

When The North Face store opened for business, the owners offered glasses of wine and live music. A hot new local band played for the customers and guests: The Grateful Dead.

The North Face's merchandising formula was novel for the times, but it worked. Interest in rock climbing, camping, and the outdoors was exploding during the 1960s. The new business made money from the start—not a fortune, but still profitable.

By then, the Tompkinses had two young daughters, Quincey and Summer (the latter so named, Susie says, because she was born on the first day of 1967's "summer of love," the zenith of the hippie movement). Neither the business nor his expanding family deterred Doug from taking off for months-long adventures—it was something he said he had to do.

Of all his adventures, one stands out—the one that he and Yvon Chouinard still talk about the most, and that seemed to mark a turning point for both men, launching them on entirely new trajectories in their careers and their lives. This was their six-month climbing, driving, and filmmaking odyssey to Patagonia in 1968. Chouinard has called it his most enjoyable adventure. Tompkins says it planted the seeds for the eco baron he was to become two decades later.

Like so many of their trips, this one began on a whim. Tompkins had come down from the Bay Area to Chouinard's Ventura beach house for a day of surfing. As they lounged on the hot sand, Chouinard mentioned a book he had just read about the first ascent in 1952 of the frigid Mount Fitz Roy on the Argentine-Chilean border. The author described the climb as his best and most difficult; Fitz Roy had been scaled only once since then.

The two friends talked idly at first about the otherworldly silhouette of Fitz Roy, and what a great trip it would be; but of course they couldn't do it—they both had young businesses in need of constant nurturing, they had obligations, they had family. But what a trip it would be. No harm in talking about it. Perhaps they could go in

July. It would take a month. No, two months. Maybe four. Who else might come?

The idle wistful talk became frenzied and before they knew it, the sun was setting and they suddenly had a real plan. They would invite three other Funhogs—the successor group to the Valley Cong—and the trip would take six months. They'd begin in July, which is winter in South America, and work their way down the coast from one hemisphere to the next, driving all the way from California, surfing on the west coast of Central America and South America, gradually moving south until they hit snow, then switching to skiing for a month in Chile. They'd feast on whatever the local cuisine had to offer and drink whatever spirits were for sale, timing their trip to arrive at the mountain as spring came to Patagonia, sometime around October. They'd film a documentary of the entire trip, turning it all into one, big tax-deductible business venture. "Plans were piled on plans, fantasy on fantasy; by nightfall we had concocted the trip of trips," Tompkins later wrote for the *American Alpine Journal*. "We were like boys who sneak into an ice cream shop to make themselves a gigantic sundae or banana split."

The ski racer and essayist Dick Dorworth and the climber, skier, and photographer Lito Tejada-Flores piled into a van with Tompkins and Chouinard in California; the British climber Chris Jones rendezvoused with the group in Peru. On July 12, 1968, the group set out for the 16,500-mile drive to Patagonia, in an old Ford Econoline van crammed with climbing gear, packs, surfboards, wet suits for cold-water surfing in Peru, skis for the volcanic slopes of Chile, camping equipment, two movie cameras, several miles of 16-mm film, and a reel-to-reel tape deck with twenty-five hours of recorded Dylan, Beatles, Grateful Dead, and their collection of 1960s psychedelic rock favorites from San Francisco's famous Fillmore Auditorium scene. Along the way, for three and a half months, they gorged on steak dinners; ran out of food and caught wild sheep to butcher and roast; dodged bandits in Cartagena, Colombia; and awoke while camping in Guatemala to find a group of soldiers surrounding them with guns drawn, demanding to see their passports, suspecting them of being from the CIA. It was a dangerous time in parts of Latin America, when revolutionaries, counterrevolutionaries, left-wing assassins, and right-wing death squads vied for supremacy, and covert American machinations were suspected everywhere—at times, justifiably so. "It was a really great dirtbag adventure," Chouinard would later chortle, but at the time, it appeared to the scruffy young men that a few of those soldiers would have happily shot them in the head at the slightest provocation.

The creaking van finally reached the Patagonian pampas, the last leg of the trip, having survived a roadside rebuilding of the engine by Tompkins and Chouinard, followed by a sixty-mile-an-hour nighttime encounter with a herd of mules that appeared suddenly before them on a pitch-black road. Then one morning they saw Mount Fitz Roy looming before them, an impossibly tall, lone mountain rising up from the greens and browns of the Patagonian steppe, at once inviting and terrifying. No wonder, Tompkins muttered, only two expeditions had ever made it to the top.

After spending a month establishing camp and making slow, halting progress up the mountain, the group had to dig an ice cave 3,000 feet from the summit and wait out blizzards and high winds for four weeks, playing cards, telling stories, and interpreting each other's dreams using Tompkins's recent reading of Freud as a guide. They were forced to retreat periodically down the slope to their base camp to fetch more provisions for their wet and cramped ice cave—supplemented by the occasional sheep roast and a desperate 100-mile drive for additional supplies.

When the weather broke at last, Tompkins set the alarm clock an hour earlier and didn't tell anyone, hoping to get his stiff, grumbling friends up and out while the clear skies held. In bitterly cold darkness they began to climb at two-thirty in the morning, using a new, untried route, reaching the cloud-shrouded summit at seven that night. They became only the third group of climbers ever to stand atop Fitz Roy, named for the captain of the famous HMS *Beagle*, on which Charles Darwin made his name as a world-class naturalist and was inspired to conceive the theory of evolution by natural selection. The Funhogs had set a new standard: Their "California Route" became the most repeated by subsequent climbers of Fitz Roy for decades to come.

After sixty days of trying to reach the top (of which only five days were spent actually climbing), the Funhogs spent all of twenty minutes on the summit. This is how climbers are: they stayed only long enough to exult, to catch their breath, and for Lito to shoot a few thousand feet of film. The joy was in getting there, not in being there—a pattern that would haunt Tompkins for many years, for ill and good. They climbed and rappelled back down in frozen darkness a short time later, for fear of being caught exposed as the next storm started rolling in. They made it down safely, slept all of the next day, then drove back to civilization, where they realized that Christmas had arrived and that it was time to celebrate.

The experience reinforced the connection Tompkins and Chouinard felt for this region even as it heightened their concerns that one of the world's signature wilderness regions might be in jeopardy. Both men had noticed some disturbing signs that the natural beauty of Patagonia they had first experienced and admired eight years earlier was not quite the same this time around—forestland had been burned or cut down, and roads had been bulldozed through sensitive habitats. Tompkins griped that areas once lit only by starlight were now dotted with the electric glow of new developments. Progress, most people would call it—but Chouinard and Tompkins were not most people.

When they returned home, Chouinard embarked on a series of changes to his business. First, he decided that his popular hard iron pitons were causing too much damage to the environment—each time they were hammered and removed, the cliffs and rock faces were altered, and the experience for the next climber was diminished. So he developed new gear—aluminum chocks and wedges that could be placed and removed in cracks and crevices without the use of hammers and without causing the rock face to crumble, yet still serve climbers' needs. He discontinued his hard steel pitons, even though they were his most profitable products, in favor of a new idea: "clean climbing." Once again, he had reinvented his sport: clean climbing quickly became the preferred method for rock and mountain climbing everywhere.

The other big change for Chouinard came a few years after the trip, though it had been planned for some time: He started a new company to sell clothing for climbers and outdoorsmen—soft, comfortable, colorful, and warm, inspired by a rugby shirt he had picked up abroad. He and Tompkins had talked about what to name this new company; he wanted something just right, as magical and evocative as the mythical Shangri-la. But then they realized it was staring them in the face all along, an obvious choice: He would call his new company Patagonia. The emblem would be a stylized drawing of Mount Fitz Roy. Patagonia, which began with four employees, was destined to become one of the most successful and environmentally responsible sportswear companies in the world. (In 2006, it had annual revenues of about $260 million.)

Tompkins also returned to the United States primed for a change. He had taken on the role of director of the documentary film about their six-month trip, and he had put together what would become a cult climbing classic, *Mountain of Storms*, which focused on the ascent of Fitz Roy, along with a longer version about the whole trip, *The Funhogs*. *Mountain of Storms* won the grand prize in 1972 at the Trento Film Festival in Italy, which convenes annually to honor the world's best mountain, adventure, and exploration films (that year the competition included films from the legendary marine explorer Jacques Cousteau and *National Geographic*). *Mountain of Storms* began making the rounds at college campuses, where it spawned something of a new American subculture of climbing and traveling—an adventure lifestyle that inspired innumerable treks around the world imitating the Funhogs. Hunter S. Thompson eventually appropriated the term to describe his style of covering political campaigns. In its own way, *Mountain of Storms* did for climbing and adventure road trips what the classic documentary *Endless Summer* did for surfing in 1966.

The mercantile world of The North Face seemed unacceptably mundane in comparison. Tompkins had climbed that retail mountain; who wanted to just hang around at the top? He griped to Susie and Chouinard that he was spending more energy talking to people about equipment than engaging in the sport itself. He realized his mistake, too late: It turned out he hated making a business out of his main hobby. Less than three years after founding The North Face, he sold it for $50,000, a tenfold return on his initial investment, and laid plans for a new career making adventure films. (The North Face would continue to grow into a popular brand of its own, changing hands several times and moving from specialty equipment to mass consumer clothes and products; it is now part of the VF Corporation's $350-million stable of brands, which includes Vans and Jansport, accounting for half the backpacks sold in the United States.)

The filmmaking career was not to be. As Tompkins finished *Mountain of Storms* and started casting about for his next project, another opportunity arose to sidetrack him: his wife's new business. When he had returned from his six-month Patagonian trip, he found that Susie had not simply waited at home with the kids. She and a friend, Jane Tise, who had worked at the ski shop before The North Face took its place, had started their own kitchen table business, the Plain Jane Dress Company. They had been making hippie-style, floral-patterned knit mini-dresses and marketing them to local stores. The office was the Tompkinses' house, an arrangement that solved the problems of child care and the fact that they couldn't afford extra rent.

They started with twelve dresses and showed them out of Susie's old station wagon. Susie Tompkins's gift, colleagues would later say, was a knack for spotting a trend before it was a trend: Everyone—

specifically the masters of the consumer economy, teenage girls—loved the dresses. They were what Susie called "Londony"—snug, short, and sexy—and no one was selling anything quite like them in America. In short order, Susie and Jane found they barely could keep up with the demand from local boutiques. They took on a third partner to handle sales and soon Plain Jane clothes were in Macy's—the promised land for any upstart clothing designer.

Doug began helping Susie part time with the business, and to his surprise, he became intrigued. Unlike The North Face, women's fashion posed no conflict between business and pleasure for him. After being apart so much, he and Susie relished working together. And though fashion is of little interest to mountain climbers—Tompkins has been dressing in the same blue jeans and polo shirts since he was sixteen—the idea of building a business, a brand, and an image, and using that to turn some colorful fabric into a profit, sucked him in. He saw Plain Jane had a bit of magic to it, but the company was disorganized, lacking a coherent vision—it needed someone to handle its business, marketing, and image. Why not him, even if this meant making it up as he went along?

He offered to invest some of his earnings from selling The North Face to become a partner, and to work for a while getting the business end squared away. Then, he figured, he could return to filmmaking. But by the end of 1970, Plain Jane had developed a line of different labels with catchy names—Rose Hips and Jasmine Tees among them—and sales had reached $1 million a year. The kitchen table and station wagon weren't cutting it anymore, and so Doug arranged the purchase of an old wine warehouse above a spice factory—the place smelled permanently of cumin and tarragon—where he set up a garment manufacturing operation and office space. The partners created an overarching company for all the separate labels, and called it Esprit de Corp. There was no turning back at that point. To Tompkins, the adventure of building an organization and pushing it to the top of an industry turned out, unexpectedly, to be as seductive in its own way as Fitz Roy had been: "After that," he recalls, "it was twenty years of a wild ride." They never expected their little family business to make them rich, he and Susie Tompkins say, but more money than they had ever dreamed of poured in.

By 1986, the company, its name shortened to Esprit and marked by the now famous, distinctive logo "E" made up of three parallel bars, had worldwide sales of $1.2 billion. Esprit was at the top of the fashion world.

Naturally, from the mountain climber's been-there-done-that point of view, reaching the summit meant it was time for a change. Or time to leave.

the empire strikes out

It's 1987: Esprit is, by all accounts, flying high. The man who once huddled for a month in an ice cave and ate half-cooked wild sheep, and who had to borrow sixty bucks from a girlfriend who wasn't sure she even liked him, now stands in his sumptuously appointed, eco-friendly San Francisco offices, dealing with another sort of survival issue: He is being courted by Helen Gurley Brown. Or rather, his marketing money is being courted by Helen Gurley Brown. The legendary editor-in-chief of *Cosmopolitan* magazine, then a staple at every supermarket checkout line in America, has come to an industrial area of San Francisco called Dogpatch, home to Esprit world headquarters, to talk Tompkins into advertising his casual California-chic fashions in her magazine.

Brown is not used to the role of supplicant, but Esprit is the label of the moment, and though she is not there with hat in hand, she is carrying her high heels in hand. At Esprit, company rules forbid spikes on the exquisite softwood floors, and so she is sliding around in her stocking feet.

"Doug," Brown says, "I know *Cosmopolitan* has an image problem vis-à-vis Esprit, and I just don't know what to do about it."

The journalist Maureen Orth just happened to be there to chronicle the exchange for *GQ* magazine.[1] One of those serendipitous moments reporters live for occurs as Brown sings her readership's praises and Tompkins all but snickers. He has been very picky about where he spends Esprit's coveted ad dollars, and he tells her the "Cosmo Girl" doesn't interest him. "Your covers don't look like our customers," he says flatly.

As Brown begins to defend her magazine's demographics, Tompkins shakes his head and explains, "It's a clash of concepts." He grabs a yellow legal pad and draws a circle; he's an inveterate diagrammer. "These are Esprit customers." He draws another circle, which barely touches the first. "These are the *Cosmo* readers. . . . I guess it comes down to a question of taste."

That Tompkins felt sufficiently secure to treat the powerful editor of a popular magazine so dismissively was a telling moment: *GQ* used it as the signature anecdote at the top of its mostly flattering story about the rise of Esprit and the man at its helm. That encounter suggested a company so sure of itself that it could blow off a cut-rate deal for the industrial-strength advertising reach of what was then the world's leading magazine for women.

But the exchange, it turned out, was more revealing of Tompkins's personality and his dedication to maintaining his brand's image than of the actual strength of the Esprit fashion empire at that moment. Unbeknownst to outsiders, including Brown, the company was in deep trouble: Nineteen years after its humble start as Plain Jane, it had expanded too rapidly, had borrowed too much, had stumbled in its latest fashion lineup, and was hobbled by increasingly fractious disagreements between Doug and Susie Tompkins about fundamental matters of image, direction, and design. The $1 billion company—

and Tompkins—had reached a turning point, as the fairy-tale story of the little hippie business and the power couple who built it into an empire hit a bone-cracking bump in the road.

Until then, Esprit had seemed immune to missteps. From the moment Doug came on board to manage the business end of Plain Jane, the 1970s had been a time of rapid expansion for the company. Esprit wasn't exactly a knockoff dressmaker then—it didn't out-and-out copy other designers—but it did follow rather than dictate fashion trends, nimbly riding the latest fads and delivering well-made, moderately priced clothes to the major department stores and shops that catered to the high school and college crowd. By the end of the 1970s, the company had $120 million in annual sales, and Susie and Doug had bought out their partners. Once Esprit was theirs, they shifted most manufacturing abroad, ruthlessly closing down their primary garment operation in San Francisco just as it was about to unionize (eventually costing them a $1.2 million judgment for back wages). They soon became the largest exporter of clothing from Hong Kong at the time.

When Esprit's headquarters (including all its inventory) burned down in 1976, Doug designed a showcase replacement building at the site of the old winery on Minnesota Street, importing 100-year-old, forty-foot recycled timbers from an old sawmill in California's gold rush country—reflecting an environmental sensibility twenty years before recycled lumber became the fashionable green alternative. The airy brick structure Tompkins designed featured skylights, a gourmet cafeteria, a running track, tennis courts, and artful decor that included an extensive collection of rare and antique Amish quilts. He and his wife also designed a unique work ethic: Although Esprit was legendary for offering relatively low salaries, it doled out unusual benefits to its workers: a 52 percent discount on Esprit clothing; money to buy tickets to the theater, ballet, and opera; river-rafting vacations underwritten by the company; free foreign-language lessons; and use of the company ski lodge in Lake Tahoe. Meetings could never have more than four people involved. At Doug's insistence, everyone flew coach, even he and Susie, as Tompkins would not tolerate the extra expense of flying first-class. Employees sometimes called the place "Camp Esprit," while less charitable observers thought it resembled a cult.

Tompkins's method as an employer—like his philosophy as a climber and a pilot—was to size people up, decide what they were capable of accomplishing (more often than not exceeding their own estimations), then persuade, push, and demand that they perform to his expectations. He was less impressed by a job candidate's experience in a specific position and more interested in reading habits, vacation plans, and ambitions. He also seriously considered making a practice of examining prospective employees' driving records to be certain they had enough temerity to break the speed limit now and then. He was easily persuaded that intelligence, fearlessness, and common sense were the most important qualifications for any job—perhaps because he had succeeded in fashion, about which he knew nothing when he started, and had designed a building that won architectural raves and awards, though he was no architect.

Peter Buckley is the quintessential example of Tompkins's hiring philosophy in action: the placing of an essentially unqualified but ingenious person in a position of great responsibility—running several of Esprit's international operations, with great success.

Buckley was earning his law degree at the University of California Hastings School of Law in San Francisco when he first met Tompkins in 1970, having been introduced by a classmate who worked part-time at Esprit. Tompkins sought someone to school him in using a trampoline, and Buckley, an all-American high diver in high school, was an expert. The easygoing blond surfer and the intense mountain-climbing executive hit it off immediately, working out a couple of times a week, using a trampoline at the University of California–Berkeley gym. Tompkins, an apt pupil, became fascinated with all sorts of acrobatic arts; he and Buckley checked out all the circuses that came to town, looking for inspiration. Finally he asked Buckley, "Hey, if I buy a trampoline, would you come over to the factory and work out with me?" Buckley said sure, and a short time later, Tompkins had cleared out a space inside Esprit headquarters, where the ceilings were very high, and set up a trampoline area in the midst of a fashion house, practicing flips and midair twirls while the latest fall frocks were assembled below.

For the twenty-three-year-old Buckley, the entrée into Tompkins's world of fashion, A-list parties, and gorgeous models proved irresistible. When he graduated from law school and opened a small practice in the Bay Area, his days paled in comparison, and he soon gave it up when Tompkins offered him a job at Esprit's new fabric operation in India. He knew nothing about the fashion business, but Tompkins assured him he would learn as he went, that all Buckley needed was common sense—assurances that were soon followed by much screaming over the international telephone when a shipment got screwed up or materials were the wrong color or monkeys ate the silk screens.

Doug was a great friend but a frequently difficult boss, Buckley says, although just when you'd get ready to throw up your hands and quit, he'd do something so incredibly kind and considerate that you'd swear lifetime loyalty to him. For Buckley, that moment came when he was summoned to Hong Kong for a meeting that could just as easily have been dealt with by phone. Buckley arrived a mess, haggard and disheveled, with Tompkins bitingly wondering what the hell was wrong with him, until he confided that his girlfriend had just left him. Tompkins's mode shifted from anger to horror to solicitousness; he began racing around the Hong Kong factory, scooping up one item after another—anything he felt the ex-girlfriend would like. Then he packed up his friend and sent him back to the airport for the next flight to India to try to get his girl back. "But what about the meeting?" Buckley asked, and Tompkins answered that it didn't matter: "This has to take priority."

Two years later, after a brief return to his mothballed law practice, Buckley agreed to take over Esprit's debt-laden operation in Germany. There, he came up with a cost-cutting plan he felt would save the failing subsidiary within the year. But when Tompkins arrived to take stock of the situation, he looked over the books and announced it was time to cut their losses and close down the European branch.

Impulsively, Buckley said, "I'll buy it. I think I can turn it around."

Doug raised his eyebrows; he knew Buckley was as broke as Esprit's German operation. "How much have you got?"

Buckley pulled out the one bill in his pocket. "I've got one German mark."

Tompkins smiled. "I'll take it." Esprit Germany was Buckley's.

A few months later, when a new low-cost line of mix-and-match tops and bottoms in just a few colors hit the stores, Buckley's desperate efforts to economize were viewed as fashion simplicity genius. Orders poured in, saving the business. Buckley eventually rejoined the mother company, selling back his thriving Esprit Germany to become a one-third partner in a newly formed Esprit International. That one German mark would make him a multimillionaire, but Tompkins never blinked, Buckley says. They laugh about it. And to Buckley, this disregard for money—not to be confused with irresponsibility toward money, for Tompkins is often frugal—is a defining characteristic. It explains why their friendship survived and grew after the turnaround in Germany. And it would seem to be a prerequisite for a man who would decide to give away so much of his wealth to environmental causes—while persuading Buckley to do the same.

Back in the United States, the Tompkinses launched a complete makeover of the entire Esprit enterprise, a roller-coaster process for fixing the brand in the public consciousness that would take from 1980 to 1985. All the old Plain Jane brands, which had different names and logos for each clothing line, were discarded; everything would be branded "Esprit."

Susie became chief of design for the company, with six young designers working for her, while Doug turned his attention toward fashioning the new brand image, with the goal of positioning Esprit as a company that no longer followed trends but instead purported to lead the fashion world with a line of youthful, sexy, casual, sporty clothes for both teens and adults. No corporate clothes. No gowns or high heels. No clothes that hinted at being old. The vivid, crayon-bright colors of the first Esprit catalogs leaped off the page; when rival houses copied the look, Esprit led the mid-1980s into more muted pastels. These were clothes that looked good, that you could wear anywhere—at parties, work, or play—but that you could do a somersault in, too, Tompkins insisted. They were clothes he and his wife would wear.

Signs were posted around the shop: "Thou Shalt Not Knock Off." They were accompanied by others instructing, "Life Is Entertainment, Survival Is a Game," and Tompkins's then-favorite credo, "No Detail Is Small." At work, he lived that latter motto as if it were a religion, right down to personally positioning the "E" logo in just the right place on a new line of kids' polo shirts, and writing the script for how Esprit receptionists had to answer the phone: with a cheery, bright, "Hi, Esprit!" It seemed a curiously obsessive aspect of his personality, given his continuing seat-of-the-pants approach to the adventurous side of his life, and his habit of still taking off for months out of the year. This was the same guy who impulsively decided to leave a new business and a two-week-old baby for Patagonia for six months; who years later abruptly decided to drop off a party guest at Yvon Chouinard's house by landing his private plane on a closed, partially constructed freeway next door; and who, on another occasion, decided it might be a good idea to fly a sea-skimming route below radar in fog, going *under* the Golden Gate Bridge, because he was too impatient to wait out the fog for official clearance. This combination of fearless impetuosity and risk taking along with attention to fine detail drove his colleagues and employees mad at times, but it served the business well—as it would his later environmental endeavors. "The product has to be good," he told his Esprit staff, "but it's the image that sells."

To construct the new image for Esprit, Tompkins interviewed numerous prominent photographers, then decided he needed one who wouldn't even come to San Francisco. He had to have Italy's Oliviero Toscani, one of the most sought-after and controversial fashion photographers in the business. Tompkins wooed him for the better part of a year, even taking Italian lessons to help make his pitch, and Toscani finally relented to the man he later likened to a Medici prince. Tompkins and Toscani decided to reject the posed artificiality of most fashion photography of the time and to focus instead on provocative images and emotional appeals. The oversize color catalogs that resulted relied on marketing by association, de-emphasizing the actual products being sold while emphasizing the lifestyle attached to the product—how wearing, using, or touching Esprit clothing was supposed to feel. The models weren't posed; there was no perfect draping of the clothing so that the fashions would be clearly visible. Instead, Toscani had people kissing, hugging, jumping, and playing in the snow—the clothing seemed an afterthought. This sort of advertising is pervasive now (one widely praised contemporary example: the Apple iPod television ads featuring compelling music and psychedelic images of dancers—but no information about, or even images of, the iPod itself). A quarter century ago, however, the approach was derided by analysts, competitors, and even Esprit's staffers. Tompkins, they said, had lost it.

But the catalogs struck a chord with customers. They called, they wrote, and they bought, setting sales records for the company. Esprit became a highly recognized brand through those catalogs—even before the billboard, television, and magazine ad campaigns that followed. The company soon branched out into children's clothing and men's and women's sportswear, and launched a "real people" campaign in which company employees and, in subsequent photo shoots, Esprit's customers, were used instead of professional models. Tompkins wrote pithy, short biographical statements for his models. ("I'm a connoisseur of fine junk food, and have been known to be an occasional food slut." "A funny thing happened to me at an Esprit catalog shooting. I fell for the photographer's assistant and decided to run away with him to Italy.") While critics again derided Tompkins's approach—it was too cutesy, too unprofessional; the models were too plain—the campaign evoked a huge public response. The company was flooded with thousands of letters thanking Esprit for showing clothes on people who looked like regular people—albeit the young and attractive regular people the Tompkinses invariably hired—rather than fashion models that no mere mortal could hope to resemble. "Women who wear Esprit are the new feminists!" Tompkins crowed in a newspaper interview at that time; it would make him wince years later.

Tompkins, intent on leading a "design revolution," leveraged Esprit's popularity in the early 1980s to force major department stores to construct a "shop within a shop" display with unique lighting, music, bags, and uniforms, where Esprit clothes, shoes, and accessories were displayed and sold in one place, instead of haphazardly throughout different departments. Only Esprit got such treatment at the time. Then the company launched a worldwide chain of freestanding Esprit stores, the first an opulent $15 million, 30,000-square-foot superstore in Los Angeles, partly designed by Tompkins in a sprawling former West Hollywood bowling alley and roller rink. In San Francisco, a huge warehouse outlet store drew hordes of shoppers and rows of tour buses on weekends. Nearby, an Esprit café and restaurant opened, but only after Doug held three meetings to choose just the right salt and pepper shakers. Then he ordered a worldwide search for the ideal doggie bags, which were so beautiful customers used them for purses.

Throughout this rapid expansion and transformation, Doug still took regular, extended trips to climb and to pursue his latest passion, white-water kayaking. His group of kayakers made a series of first descents—the first successful passages—of some of California's most rapid, turbulent rivers. When he wasn't on an expedition, he was happiest in his plane, flying to Mexico or Ventura to see Chouinard. The "vacuous" world of fashion—where he never really fit in and made few lasting friends besides Toscani—was losing its luster for him, even as he tried to revolutionize it. Instead, he lived for his escapes to nature. "I felt best when sleeping under the stars, in forests on high vertical walls, on glaciers, or along great steep ravines," he later reflected. Between his adventuring and business travel, and Susie's own extensive travel in pursuit of new fabrics and design trends, the two spent barely four months of the year together. Publicly they still spoke of the joys of working together, but the distance between them became more than a matter of separate travels. Friction about the direction of the company, and the direction of their lives, was beginning to take hold, too.

Each of them was feeling an increasing disconnect between the global enterprise of Esprit on the one hand, and on the other, their love of the outdoors, the simpler fun of their first small businesses, and the simpler lives they had led before Esprit grew so much and so fast. Big fashion had made them rich, had given them choices in life that most people would envy. Yet they were environmentalists and iconoclasts at heart and their sensibilities were more in tune with counterculture values than with corporate and consumer America. Esprit was the epitome of corporate success and retail consumption: new looks manufactured every month or two, a global enterprise that in many ways defines what is most ephemeral and least sustainable in the modern world. "We're manufacturing desires to get people to buy products they don't really need," Tompkins fretted. Even in his outdoor adventures and hobbies he started to feel he had come up short—that he was treating nature like an outdoor gymnasium or amusement park, a setting to be exploited rather than a resource to be treasured and preserved. As their friend Yvon Chouinard sought to do at Patagonia, the Tompkinses began to seek ways to make their business more a force for good environmentally and socially, with bold plans to export *to* China, to use more sustainable manufacturing processes, to promote healthy lifestyles, and to publish books (including Doug Tompkins's reflections on his and Esprit's philosophy, and his disdain for the so-called Reagan revolution—what he called "those bozos running the government with their hundreds of billions worth of deficit").

First came one of the nation's earliest corporate AIDS awareness efforts, beginning with a full-page ad and an open letter in their 1985 catalog condemning bigotry toward AIDS victims and inaction by the Reagan administration, and explaining at length that AIDS was not a "gay" disease. Using a fashion catalog to bring such concerns into public view was unheard of at the time, but it broke a barrier, leading to a new niche of progressive causes married to corporate profits—the niche of "companies that are trying." For the rest of the decade, the Esprit catalogs also became a vehicle for discussing social causes, from homelessness to recycling.

Esprit next created an "eco desk" where staffers worked on finding more ecologically friendly means of manufacturing for the company, while coordinating environmental volunteer opportunities for Esprit employees. Staffers would be paid up to ten hours a week for a like amount of volunteer work at a nonprofit. Guest lecturers from the conservation community were featured weekly at Esprit headquarters, including Dave Foreman—a cofounder of the radical organization Earth First!—and Jerry Mander, a former top advertising executive turned environmentalist, critic of the global economy (and companies like Esprit), and author of the book *Four Arguments for the Elimination of Television* (1978), an attack on the industry he once helped lead.

Around this time, in the mid 1980s, Doug Tompkins began spending increasing portions of his workday on environmental issues. He'd come in early in the morning and labor through lunch on his pet projects, then work on Esprit business until well into the evening, sometimes until nine at night, just to get everything finished. At home he began devouring books on ecology, the media, and philosophy, doing the homework and study that he had given up so long ago when he ditched high school. Mander's anti-television book particularly shook his self-image as Esprit's benevolent "image director." According to Mander, there was nothing benevolent about the advertising industry's use of imagery—it was a form of "Orwellian mind control" programming Americans to consume. The average American could name far more brand names than plant names, Mander pointed out. With an uncomfortable shock of recognition, Tompkins concluded that this description closely paralleled his own use of imagery to establish Esprit as the brand people had to have in their closets.

In short order, Tompkins launched a new public service campaign in the Esprit catalog called, "Buy Only What You Need," which counseled customers to buy fewer clothes from Esprit and elsewhere, and keep what they owned longer. The consumer culture was laying waste to nature unnecessarily, as Tompkins saw it. Why not sell clothing designed to endure longer than a season or a passing fad? Special tags were put on the new lines of clothes warning customer to "Buy Only What You Need." His colleagues and investors went ballistic. What kind of company, they wondered, urged its customers not to buy its fashions? That may be a more sustainable lifestyle, but it doesn't make for a more sustainable *business*, they argued.[2] But Tompkins stayed firm, and his idea seemed to pay off, at least for a while: The new line of clothes made headlines and money.

His determination to make even bigger changes was driven by two successive events: a flight in his private plane over western Canada, and his reading of a book by the philosophy professor George Sessions and the ecologist Bill Devall, *Deep Ecology: Living As If Nature Mattered* (1985), which he said produced in him a "powerful epiphany." The book laid out the principles of a relatively new philosophy

called deep ecology, developed in the 1970s by a Norwegian philosopher, activist, and (particularly appealing to Tompkins) mountain
climber, Arne Naess. Naess argued that nature has intrinsic value,
independent of how useful or attractive it might be to humans, and
that therefore no form of life, human or nonhuman, has any more or
less value than another form of life. In short, Naess urged a biocentric
worldview, rather than an anthropocentric mind-set, and this meant
humans had no right to destroy nature or biodiversity except to meet
vital needs. Gas-guzzling cars, suburban sprawl, and walk-in closets
full of Esprit fashions did not qualify as "vital needs."

Tompkins distilled this message to a simple comparison. The old
way of thinking was that if something was good for humanity, it was
good for the world. The deep ecology way of thinking flips that for-
mulation: If it's good for the world, it's good for humanity. To some,
this is a radical and dangerous idea that threatens the very fabric
of human society, religion, and economy; to Tompkins, it seemed
so simple and so obvious that he couldn't believe he hadn't seen it
before—or that more people weren't flocking to its life-affirming
principles.

The second event was more visceral than academic, a flight over
British Columbia, where he goggled at the stark vision of mile after
mile of clear-cut forest beneath him where once the landscape had
been a gorgeous carpet of green. Felled trees stretched as far as he
could see. How could people cause such destruction, so dramatic that
it looked like the site of a nuclear detonation? And how could he not
try to do something about it? This devastation of irreplaceable nature
was the logical result when a society rejected deep ecology's ideals,
Tompkins decided with a jolt. This was what the prevailing philosophy of "wise use" of the land permitted, because, it was argued, not
all the trees were taken down—a few were left standing to satisfy environmentalists, in the sort of compromise that was always made to
justify what Tompkins could only describe as carnage. The sight left
him in a "green rage," he says, primed to change the world and him-

self. And he knew he could never accomplish the sort of change he craved from within Esprit.

"My zeal for business began to fade away," Tompkins later wrote of his attitude in the mid-1980s. "I had realized that the production and promotion of consumer products not vital to anyone's needs were as much a part of the eco-social crisis as anything. I was, simply, contributing to the problem itself. I had to do something else. I had to figure out how to get on the solution side of things."[3]

He turned to Jerry Mander for help. As a young ad man in 1964 Mander helped design a campaign for the Sierra Club's legendary first executive director, David Brower. Brower wanted to halt seemingly unstoppable plans by the federal government to build hydroelectric dams that would flood the Grand Canyon, turning mile after mile of river gorges into mile-deep lakes. Today such an idea seems laughable, an abomination no one would tolerate. Yet then, the dams were portrayed as essential, a sign of progress in the modern world, and those opposed to them as tree-hugging fools and Luddites. Mander seized on one of the arguments developers had advanced in favor of the dams: that the resulting lakes would rise so high that power boaters could cruise near the canyon rim walls for a better look at the parts that weren't submerged. What followed made advertising history: a series of full-page ads in *The New York Times*, the first of which asked: "Should we also flood the Sistine Chapel, so tourists can get nearer the ceiling?" The public outcry was so swift and so immediate that the Grand Canyon dam project soon died. So did Mander's enthusiasm for using the power of such imagery to sell goods. He quit his successful advertising business and founded the Public Media Center in San Francisco, a nonprofit that worked only for environmental, social, and antiwar causes. In grudging admiration, the *Wall Street Journal* later called Mander the "Ralph Nader of advertising."

Tompkins started making donations to a variety of the center's environmentalist clients and public service messages. He joined Chouinard to rescue a forest in Chile, then donated funds to support additional land purchases. But he wanted to do more, to have more of an impact, and he asked Mander to help him plan an environmental think tank. As they talked it over, Tompkins realized that there was no shortage of thoughts and ideas on what needed to be done about pollution, extinction, and conservation. Grand ideas and reports from think tanks for saving the world had been gathering dust for years, or had been cut, rather than supported, by a series of antienvironmentalist administrations in Washington. The problem wasn't a lack of ideas; it was a shortage of cash and people to make those ideas a reality. So instead of a think tank, he and Mander started planning a foundation to promote projects that reflected the ideals of deep ecology. The one thing missing was the money to make it all work—they'd need a big pile of it. Tompkins quietly awaited the right moment to cash out of Esprit.

It came in 1988. The year before, Esprit's revenues had dropped precipitously. Tastes had shifted, and Esprit's styles suddenly weren't moving as they used to. The department stores that had been coerced into constructing the special shops within shops were in no mood to be a friend in need. The company had gone $75 million in debt to construct fourteen lavish stores of its own, and business was not meeting expectations. The Esprit computer system was aged, and customer service for phone orders was a mess—two areas in which neither Doug nor Susie took a personal interest. Annual profits fell from $62 million to $10 million. The Tompkinses were forced to cut a third of the U.S. staff, to 1,500. During one ugly disagreement, Doug tried and failed to fire Susie as the chief designer while she was away in Hong Kong. She foiled the attempt by refusing to come to the phone; Doug eventually cooled off, for the time being.

Despite the drop in earnings, the brand still had enormous cachet, making it attractive to potential buyers, and continued to flourish in Europe. A two-year tug-of-war ensued, in which the company devolved into bitter factions, as Susie and Doug fought over everything and rival camps sprang up within the staff. No detail was too small to battle over. Anything from the color of a blouse to the size of an advertisement could become an occasion for conflict. Susie wanted to market to a more mature, corporate customer, while Doug insisted the company must never abandon its appeal to youth. They moved into separate houses and filed for divorce. The *Wall Street Journal* published a front-page story on the tumult at Esprit that depicted a poisonous atmosphere and shattered morale. "We've been fighting like cats in a bag for 15 years," Tompkins told the *Journal*. "It's just a matter of convenience that we remain married."

Lawsuits followed, a new board of directors was seated, and a CEO not named Tompkins was appointed to take day-to-day control of Esprit for the first time in two decades. For a while Doug seemed to have come out on top, holding more sway with the board, while Susie lost her duties as chief designer and was relegated to a consultant role with diminished power. The company went into recovery mode while a buyer was sought.

By 1990, Esprit was up for auction. Both Tompkinses continued jockeying for control and advantage, but at the last minute, before any bids were made, Doug changed course and offered to sell his 50 percent of the U.S. division of the Esprit company to Susie. Then he sold out his one-third interest in Esprit International. It seemed Susie was the victor after all, winning control of the company at last. But Esprit faltered badly in the next two years, and she sold her interest as well, moving on to remarry, to pursue social and environmental philanthropy, and to become a major activist and campaign donor to Democratic candidates. She is a close friend of Hillary Clinton and a force in California politics, part of the rarefied upper atmosphere of Bay Area society—where Doug once could have gone as CEO of Esprit, but never wanted to go.

Esprit, meanwhile, has been transformed into a shadow of what it once was; its ownership and primary business are now outside the United States. Only the distinctive logo remains. Caffe Esprit and the massive outlet store are long closed. The superstore in West Hollywood stood vacant for a decade and eventually was transformed by layers of plasterboard and linoleum into a chain drugstore. Esprit's beautiful headquarters in San Francisco, with its huge, century-old, forty-foot beams, has been taken down to make way for a condo complex.

Doug Tompkins, it turns out, walked away the big winner. When the dust from his multiple sales settled, he had $200 million in the bank. And he was itching to give it away.

lost and found

*WE ARE THE FIRST generation in 100,000 generations of
human evolution to have our lives shaped—not by nature—but by an
electronic mass media environment of our own making.*

*LIKE CAGED ANIMALS we have lost our bearings. Our attention
spans are flickering near zero, our imaginations are giving out,
and we are unable to remember the past.*

—SIGN HUNG BY DOUG TOMPKINS IN THE
NEWLY OPENED FRONT OFFICE OF HIS
PRIVATE PUMALIN PARK IN CHILE

Kris McDivitt found her dream job in high school. And though it took twenty years, that job eventually brought her to Reñihue Fjord, to the conservation project of a lifetime, and to Doug Tompkins.

A California girl, Kris grew up on her grandfather's ranch south of Santa Barbara. But the family had a beach house in Ventura, too, which is where she met a mountain-climbing surfer—Yvon Chouinard, who lived next door. He gave Kris summer jobs at his climbing equipment company, and when she was done with ski racing and college in Idaho, he hired her full-time.

That was in 1972. She grew up with the Chouinards, saw the birth of the Patagonia company when it had just her and four others on the staff, worked feverishly to make it succeed. And she took pride in the strong environmental commitment that has made Patagonia such a standout among American businesses, dating all the way back to her first full year on the job, when the company gave up scarce space in its cramped offices to a young activist working—successfully, it turned out—to restore the polluted Ventura River and bring back the steelhead trout that spawned there. That commitment continued with the costly decision to use only organic, sustainable cotton in Patagonia clothes, and with the company's decision to donate 1 percent of its sales every year to environmental causes—and to champion a similar program, One Percent for the Planet, for businesses nationwide.

Kris is athletic and graceful, bluntly outspoken when necessary, but warm and diplomatic by nature, with a fondness for quirky, colloquial phrases. (When something is odd or unreliable, in Kris-speak it's "funky monkey.") She was the natural choice in 1988 when Patagonia needed a new CEO, the high school part-timer rising to the top, coolly competent, trusted like family, and always ready to make sure the wheels stayed on when Chouinard went off on one of his trips with Doug Tompkins. By the time she hit her twentieth year at Patagonia, it seemed she was the only staff member who had not traveled to the company's namesake region in South America. In 1991, the Chouinards insisted she finally accompany them on a trip down to southern Argentine Patagonia. She was newly divorced, ready for something new, and she agreed to come.

The visit proved to be a life-changer. It wasn't the forest that drew her, as it had Tompkins. It was the grasslands of Argentina, the vast sea of browns and greens, wind moving through it like the waves that lapped the Pacific beaches of her youth, the air crisp, the snowcapped Andes towering over everything, distant but commanding. She had been riding on a bus, staring out the window, and suddenly she was on her feet, saying, "Por favor, señor. Please let me off." She still had a couple of miles to go, but she just wanted to walk, to let the wind and the vastness wash over her as the bus vanished and the solitude settled around her. She knew at last why Yvon loved this place, felt its pull. It was the old West, the way it must have looked and felt two centuries ago, powerful, alive, untamed. The road was a puny thread in that vastness, a meager scribble in a place that still outshone man's reach. By the time she reached town, she was thinking, *Some day I'm going to live here*. She didn't know how, or whether it was practical, but she identified with that landscape in some profound way she didn't quite understand, and didn't really have to. "It was love at first sight," she says. "I wasn't really contemplating leaving my work at that point. It hadn't occurred to me. But I felt something was coming, something would materialize."

Something did, or rather some*one* did. She had gone to meet the Chouinards for dinner. "It was a restaurant in some little funky monkey town; all I knew was that Yvon was meeting someone there. We were having drinks and in walks old Tompkins."

They'd known each other for years—you couldn't be around Chouinard for two decades and not know Tompkins—but it had been quite some time since they chatted. They had led very different lives, in different towns, with different ambitions. Now something changed; something had put them on the same page. Tompkins had come to pick up Chouinard so they could go climb something, but instead he lingered and talked with Kris about his life here, his conservation work, his plans for the land he had bought and come to love. She realized he had years of work mapped out for himself—decades, really. And she was mesmerized. It seemed so prescient to her, so fateful for her to have this emotional reaction to the landscape, then to meet up with this passionate, committed man, this person she considered a living legend, who had the very same connection to this place that she was feeling.

Kris had come for a ten-day visit. She stayed five weeks. Then she returned to California and broke the news to the Chouinards: She was retiring. And she was marrying Doug Tompkins and moving to Patagonia. No one, not even Kris, could figure out which of those three developments was the most surprising.

She had fallen in love in her early forties with this man, so seemingly gruff, distant, and guarded, yet passionate in his commitment, in his desire to do something really good, to lead something really good. And she saw his joy in finding someone who wanted to share that, to help build it, and to take it in new directions, he to the forests, she to the grasslands. They moved into Doug's home in Reñihue, by then a beautifully restored wooden ranch house, with no electricity, no phone, no appliances, no television, not even a refrigerator. Doug was living what he preached: He wrote letters longhand and sat by the fire; and in the kitchen, an old-fashioned icebox was kept cool by a vent that circulated chilly air from beneath the house. Ice cream became a rare and special treat you went out for, which is what it had been before World War II. Kris found this different and liberating—she saw that her life had been filled with stuff she didn't need and didn't miss. It felt right.

The timing of Kris's arrival was another matter. She got there right about the time that the Chilean press, politicians, and public became fully aware of Tompkins and his large land purchases in Patagonia—and of the fact that they might not completely approve of what he was doing. Doug had traded obscurity for notoriety by lodging a formal complaint about salmon farms polluting his property and the surrounding ocean, killing enormous amounts of marine life. Salmon farming is one of the crown jewels of the Chilean economy, part of the "Chilean miracle" once touted by President Reagan and advocates of "Reaganomics" as evidence that their ideas about privatization and unregulated market economies were sound. Criticizing the salmon industry was considered bad form and impolitic, especially when the source was a gringo environmentalist millionaire who, the Chilean press suddenly realized, had bought up so much land he had become the second largest landowner in Chile—so much land, in fact, that his holdings effectively sliced Chile in half. It didn't matter that he was right, that salmon farming really was less a miracle than an ecological disaster. In fact, his being right made things worse. A delegation of Chilean politicians responded in short order, appearing at Tompkins's gates to begin an investigation. But the investigation was not of the salmon farm and its alleged pollution.

They would be investigating Tompkins.

The funny thing about wanting to give away a boatload of money, Doug Tompkins says, is that people either think you're up to something nefarious or think you're crazy. Or both.

This is what nearly always happens to eco-philanthropists—it's been the same for the past 100 years, he says in his gravelly, deliberate voice, staring ahead but looking inward, resignation written in his expression, arms crossed. He fully expected opposition and acrimony in Chile, if not ferocity. You'd be crazy not to anticipate some kind of trouble, he asserts, given the stormy history of America's early national parks, each of which was opposed vehemently and embraced only much later on. But Tompkins's calmly historical perspective on the inevitable travails of eco-philanthropists runs contrary to the most common portrait of him in the press. The same story seems to get recycled every year or two in the U.S. media. It pegs him as being caught totally off guard by the controversy that has swirled about him for decades, a naive true believer who couldn't see the hornets' nest until he was being swarmed. He laughs at it now, if a bit wearily. Still, if not surprised at having his environmental work attacked, he nevertheless finds the reactions to his philanthropy at home and abroad both perplexing and saddening to ponder.

Think about it, he suggests: When he used his money and clout to ship American jobs overseas, or to strong-arm department stores into giving his brand preferential display, or to expend huge amounts of energy and resources selling products nobody needed, he was arguably doing considerable harm to nature and humanity, yet he was lauded as a visionary entrepreneur, a properly hard-nosed businessman, and he was richly rewarded. This was capitalism at its respectable best, doing what it was supposed to do: maximize profits, any way necessary. And if he had done the same thing in Chile, if he used his Esprit fortune to snap up Patagonian forestland, pastures, and coastal properties in order to mine, dam, or develop them, he would have been welcomed with open arms, no questions asked, as countless other foreign investors had been welcomed into Latin America's most wide-open free-market economy. That was, after all, what all the old *Pinochetistas* had been hoping to do before their dictator's downfall. He would be living the dream.

Instead, he bought the broken-down ranch and the surrounding land at Reñihue to preserve it as a world-class national park that would one day be donated to the people of Chile. He stopped the logging on his lands, canceled development projects, and moved to protect and re-store native species that generated no profit, in place of the non-native cows and sheep that profitably devastated the fragile landscape.

Meanwhile, in the United States, he parceled out grants to little-known environmental organizations that showed a willingness to scrap and offend in the name of saving wilderness, species, and the environment. Tompkins became a new kind of green sugar daddy: He supported dozens and eventually hundreds of activists and organiza-tions that had not been considered by the major grant-issuing foun-dations, many of which would rather support more established and less confrontational conservation groups, the sort that did substantial good works but that rarely caused businessmen and politicians to rip their hair out. Tompkins wanted more hair ripping. And so his foun-dation gave important support to fledgling groups that were attack-ing logging and development in important wilderness areas by suing under the Endangered Species Act and other powerful federal laws dating from the 1970s, laws that government and industry had been flouting for years. A number of those groups have had stunning suc-cess, preserving millions of acres of wildlands and forcing polluters and developers to change their ways; a number of these groups might not have made it without Tompkins's grants.

In short, Tompkins did things with his money that indisputably benefited nature and humanity far more than he ever did at Esprit. And for that he has been vilified.

He is not the least bit apologetic for what he has done and whom he supports—he takes pride in both. As an environmentalist, he is as forceful, stern, obsessive, and convinced of his own righteousness as he was as a fashion mogul. He is a harsh critic of the global economy, of business as usual, and he is quick to take others to task for not taking a stand against the destruction of nature—or for remaining willfully blind to it.

"Nature is collapsing; science is telling us this: The world is collapsing," he says. "But it's more than science—you can see it with your own eyes. It doesn't take much, if you travel around and see. Beauty is the baseline. Nature is beautiful on its own. But if you open your eyes, you can see how ugly the world has gotten from the technological industrial society. You can see it—climate, extinctions, fish kills, ocean dead zones, diseased forests. Just connect the dots. The world is collapsing. But people won't connect the dots. They don't want to."

At the same time Tompkins began buying up land in Patagonia, a logging company from Washington state, Trillium, bought up 625,000 acres of gorgeous forest to the south, in Tierra del Fuego, at the tip of the Americas. Trillium bought up almost as much land as Tompkins had accumulated at the time. "They want to cut down all the trees on their land," Tompkins observed. "I want to preserve mine forever. And I'm the one threatened with being run out of the country."

When he and Kris had their fateful meeting in Argentina, Tompkins was still splitting his time between California, where he had just formed the Foundation for Deep Ecology, and Patagonia, where he was in the thick of his first phase of land acquisitions, for which he formed another nonprofit entity, the Conservation Land Trust.

He had gotten his feet wet with the initial forest land purchase and donation he had made with Chouinard in 1989—the 1,186-acre

Cani Araucaria Sanctuary, which they undertook at the behest of an environmental group, Ancient Forests International. The forest, surrounding a network of deep-blue lagoons in the collapsed caldera of a volcano, lies just outside the resort town of Pucon. It's home to rare old-growth stands of the ancient araucaria—the monkey puzzle tree, with its single tuft of foliage atop a tall, silver-barked trunk. The land became the first officially designated private park in Chile, donated to a Chilean foundation under the supervision of one of the country's leading botanists, Adrianna Hoffman. She later became head of Chile's version of the Environmental Protection Agency, and an important ally to the Tompkinses when few prominent Chileans would stand with them.

A few months after the Cani land purchase, a phone call interrupted a meeting at the Foundation for Deep Ecology's offices in Sausalito, California, where Tompkins and his staff were reviewing grant applications. The Cat Survival Trust had declared an "environmental emergency"—a desperate search for a quick infusion of cash to prevent the logging of irreplaceable South American rain forest habitat. The trust, based in Hertsfordshire, England, wanted to create a 10,000-acre ecological preserve in the Misiones province of Argentina to fulfill the group's core mission—conserving and, where possible, "re-wilding" endangered wildcats. The area it was negotiating to buy consisted of lush subtropical rain forest north of Patagonia, occupying a narrow thumb of land at the border with Paraguay and Brazil. The last of the area's once plentiful jaguar, the king of South American predators, had been killed ninety years before, but the jungles of Misiones still held small, endangered populations of other big cats: the jaguarundi, ocelot, margay, puma, and tiger cat—half the wildcat species on the continent. And wildlife biologists considered the region a prime candidate for reintroducing the jaguar into its native habitat.

The presence of these healthy populations of wildcats made the Misiones property particularly vital for environmentalists and the

Cat Survival Trust: Wildlife biologists had recently discovered that the health of ecosystems was related to—and dependent on—the health and numbers of major predators. Even populations of prey animals eventually suffered from the wildcats' absence because, with no predators to winnow their numbers, the prey would overpopulate an area and exhaust its food supply, throwing an ecosystem's entire food chain out of balance. This pattern had been seen throughout the world: The steady extermination of wolves, bears, and other large predators in the American West had led to overpopulations of deer and other prey animals in some areas, with disastrous ecological results. This property in Misiones offered a rare opportunity to preserve a rain forest ecosystem from the top of the food chain down. But now those plans were collapsing. A skittish seller, tired of waiting for the British charity to come up with the cash, had turned to a large timber company, which was eager to harvest the property. The habitats would be destroyed.

Tompkins agreed that day to provide the money to buy the 10,000 acres out from under the timber company, and El Piñalito Provincial Park was born. After the property was purchased, other rare and endangered animals and plants were discovered there: a threatened bird species never before found in Argentina, two new species of edible fruit, four new orchids, and the largest population of giant tree ferns in the country. Giant tree ferns have been driven to near-extinction because they are highly prized for making plant pots to cultivate orchids and decorative ferns.

El Piñalito eventually become the key segment in a "green corridor" designated by the environmentally progressive Misiones provincial government—a wildlife corridor 125 miles long, where fences, logging, mining, and development are banned. The corridor links the park to other public lands, allowing unimpeded migration, predation, and natural animal movements to resume for the first time in centuries. This corridor, Tompkins came to believe, is a model for other countries, including the United States, where he would help launch and fund a similar effort called the Wildlands Project. Misiones is now home to the largest intact southern Atlantic rain forest, attracting biologists from all over the world to conduct research in one of the most humid environments on earth—there is, literally, no dry season there.

El Piñalito is protected by one warden and a green-minded neighbor. They try to keep poachers in check, but even with laws protecting the green corridor's flora and fauna, illegal logging and hunting continue, and the fate of the South Atlantic rain forest remains uncertain: More than 90 percent of its once continent-wide, international reach has been destroyed during the past century. But Tompkins's landholdings have provided a check on the destruction.

Tompkins, the incessant control freak who spent three meetings fretting over which salt and pepper shakers were right for his new Caffe Esprit, could not have tolerated this lack of control over the fate of El Piñalito if it had been his project alone. Although he was happy to help such a worthy preservation effort, he decided that his future conservation plans would be focused on lands and designs over which he could exert more complete control—be it choosing which whole forest to preserve, or choosing which font to use on a hiking trail sign. In that sense, the fashion world and the environmentalist world seemed very similar to Doug Tompkins: No detail is small. His next and biggest project would be a reflection of his uncompromising aesthetic—on a collision course with the very different sensibilities of a newly prosperous Chile, where economic development had become a virtual religion and the tradition of conservation philanthropy barely existed outside Tompkins's property line.

Pumalin Park, named for the area's native pumas, started with the 18,000 acres he purchased in the Reñihue valley, that beautiful broken-down ranch with the stunning view of the fjord and its impossibly blue waters, and a snowcapped volcano framed by the front window of the ranch house. He had a homestead reachable only by boat or plane: paradise to a man who felt most alive bivouacking on an icy ledge or stretched out beneath a velvet night sky, stars wheeling overhead, no city lights to dim the view of the galactic canvas. Next Tompkins added another 24,000 adjacent acres, at first with the simple idea of setting it all aside and keeping it safe from logging and mineral exploitation—a tiny slice of Patagonia he could keep pristine forever, the perfect retirement spot.

But his ambitions became larger and grander, as in every one of his earlier endeavors, from climbing and kayaking to fashion. There had always been the drive to expand: new markets, stores, restaurants. Why not buy more land, save more land? Why not build a world-class park, unlike any in the country—any in the world? He began sketching what his ideal nature preserve would look like: the trails, the infrastructure, the nursery for endangered plants. He dreamed of creating model organic farms on the periphery of the preserve to serve three purposes—to teach the locals sustainable agriculture and thus discourage poaching and slash-and-burn growing, to make the park's food supply self-sustaining, and to use the farms as ranger stations and wardens' quarters. The park would be as environmentally benign as he could make it. Tompkins had been to parks all over the world—the good, the bad, and the indifferent—and every one of them had some flaw, some oversight, some bureaucratic bungling that ruined the experience, damaged nature, or cluttered vistas that should have been naked and free of man's imprint. Doug Tompkins knew he could do better, and his mixture of altruism, ego, and ambition produced Pumalin Park, both model and monument.

He began snapping up one broken-down farm and ranch after another, sad relics of a grander past, now gone to seed. The crassest of farming practices had been allowed on many of the properties—whole swaths of forests had been burned to make room for crops or grazing. Sometimes the fires would burn out of control and consume immense tracts; the vivid, blackened scars were still visible. Overgrazing badly damaged the lowlands; topsoil was reduced to dust in the summer and to barren mud when the weather was wet, as it often was: Reñihue gets eighteen feet of rain a year. (By contrast, the United States' wettest forest, at Olympic National Park, gets only half that much rainfall; and the wettest U.S. city—Mobile, not Seattle, which actually isn't even close—gets five and a half feet of rain in an average year.)

Much of the land Tompkins bought had already changed hands in a wave of speculation during the mid- to late 1970s, when the dictator Augusto Pinochet spoke of developing the mostly wild and unsettled south. He vowed to build a road from the Chilean port city of Puerto Montt to the very end of the world at Tierra del Fuego, promising new settlements, industry, hydropower—and plenty of opportunities to make money. The Carretera Austral—the Southern Highway—had long been a national dream, a sign of modernity and man's victory over nature, in a country that Pinochet had taken on a headlong course into privatization, foreign investment, and supply-side economic reform. But the terrain proved immensely difficult, the road proved enormously expensive, and several coastal portions of the Carretera Austral were never built. Three main legs of the route remained covered by ferry trips lasting as long as twelve hours, instead of by asphalt. Tompkins's Reñihue property was adjacent to one such leg, a sixty-mile gap where the terrain was the most rugged and the span between the ocean and the mountainous border with Argentina was narrowest.

When Pinochet's promised road and accompanying development never materialized in the remote and sparsely populated region, tenant ranchers and farmers ended up occupying most of the larger investment properties, while a scattering of small plots supported a relatively few hardy souls scraping out a living in dirt-floored poverty. Tompkins began flying his Cessna over the area, looking for parcels to buy, landing on broad pastures to chat with the locals, sometimes making an offer on the spot if the owner was present. At home, a large wall map marked his growing holdings. He bought another 75,000 acres, followed by a huge purchase of a 445,000-acre holding from foreign owners, and additional buys from Chilean absentee owners.

He also collected dozens of smaller properties, measured in hundreds of acres or less, from subsistence farmers and squatters. Even if some of the small landowners didn't have legal title, Tompkins figured it was easiest just to offer them all money, or to offer a land swap to areas more suited to farming outside his planned park.

To help with this task, Tompkins hired a young river guide he knew from his white-water rafting adventures, Jib Ellison, who would paddle out to remote plots of land that could be reached only by river travel, then negotiate deals. A few years earlier, Ellison had led Tompkins, Tom Brokaw, Patagonia's Yvon Chouinard, the writer and former Green Beret medic Doug Peacock, and the climber and filmmaker Rick Ridgeway—a group of outdoorsmen who called themselves the Do Boys, successors to the Funhogs—on a wild float through the vast Bikin River watershed in the Russian Far East. They reached the area only after one of the Do Boys bribed an Aeroflot helicopter crew to fly four hours off-course, then drop them in the wilderness, where tribesmen fed them a stewpot of moose liver and steak in exchange for Brokaw's fifth of Scotch before sending them onward.

On the strength of being able to lead that bunch safely through the steppe, Ellison became Tompkins's jungle emissary. Ellison, in turn, was steeped in the visionary zeal of Doug and Kris Tompkins. The eco baron's influence cannot be underestimated: Two years later, Ellison returned to the United States, where he eventually started his own green consultancy, BluSkye, whose first client was the biggest company in the world, Wal-Mart. Ellison, the river guide who helped Doug Tompkins save pieces of the Patagonian rain forest, became one of the main forces behind Wal-Mart's groundbreaking initiative to go green, an effort aimed at making the big box retailer, the products it sells, and its thousands of suppliers more sustainable while lessening their contribution to global warming.

By the time Ellison left, Tompkins had accumulated more than 700,000 acres, enough to make a park as big, as varied, and as spectacular as Yosemite—perhaps more spectacular, a place of snowcapped mountains, volcanoes, more than seventy lakes, rivers meandering and torrential, waterfalls, cliffs, dense forests, rocky seacoasts, and rookeries of sea lions, seals, and penguins. Each new trail, each new journey through their latest purchase, brought something else to take Doug's and Kris's breath away.

Few of the small landowners and squatters who lived in the area were entranced with this vision of a grand park for Patagonia, but they nevertheless sold to Tompkins, happy to get out of their rainy, difficult, marginal existence. A minority would not sell, however. Some of them were unwilling to give up their familiar lives and land; others were suspicious of Tompkins's offers; still others felt threatened. Some thought he was trying to cheat them somehow, because this was Chile, where 40 percent of the people then lived below the poverty line and even when Pinochet left the presidency in 1990, the old elites who supported him still had power. The locals felt as a matter of history and principle that rich people were always trying to cheat them. Some were willing to sell at first, when they assumed Tompkins was going to cut down the forest or build something real, but when he spoke of leaving everything to nature and then giving the land to the Chilean people, they started thinking he was a liar or crazy or had some other hidden plan, because in their experience no one gave away something for nothing. As Tompkins began to assert new controls over his land—stopping the slash-and-burn farming practices, chasing away poachers and illegal loggers, ending illegal but long-tolerated grazing—resentment of the new gringo land baron and his tree-hugging ways began to simmer. Where Tompkins saw in the forests and fjords of Patagonia beauty under siege and in need of rescue, his neighbors often saw the rain forest as their enemy, a relentless foe that had to be beaten back and used up as a path to prosperity and a matter of survival. It had always been that way, it was the history of man, and few had any interest in joining Tompkins in his crusade against that ancient aspect of human behavior.

His critics would later complain that Tompkins operated in secret,

gobbling up land without revealing his plans, and this charge, ampli-
fied in Chilean media reports, helped spawn more resentment and
doubts about his motives, as well as suspicion of what was always
described as his "radical" philosophy of deep ecology. His philosophy
was not said to be green, liberal, progressive, or even extreme; it was
said to be "radical," a term equivalent to "subversive" in Chile. There
were times under Pinochet's dictatorship—which had only just ended
when Tompkins made his first purchase at Reñihue—when a Chilean
deemed politically subversive faced a potential prison sentence, or
worse. This was not lighthearted or constructive criticism.

Although the perception of secrecy and hidden motives persists to
this day, attaining the status of conventional wisdom, it seems Tompkins did not operate all that quietly after the initial land purchase of
his homestead ranch. He admits he did hope to avoid driving up real
estate prices by striking as many deals as he could as quickly as he
could. And he kept his plans for the park vague at first, because the
notion of a foreign land investor choosing not to exploit natural resources was unprecedented in Chile at the time. But when Tompkins
started buying up large tracts of land in 1991, he met with the Chilean minister of land management and the minister's staff to explain
his interest in conservation in Patagonia. Tompkins received explicit
permission (though none was legally required at the time) to invest
$25 million in land and forest preservation. He also met with the gov-
ernor of Palena Province, where Reñihue (and the future park) were
located, to explain his interest in preserving the ancient *alerce* forests
and the surrounding lands. The following year, Tompkins appeared
on national television on a prime-time news show to explain his plans
in even greater depth. He has cited this early openness about his plans
many times as evidence of his "transparency," but the information
has had little or no effect on public opinion or on the media's narra-
tives that portrayed his purchases as quiet, secret, or part of a hidden
agenda.

Kris, meanwhile, had begun organizing the small community of

workers who had gradually been assembling at Reñihue to create the
new park. There were teachers to hire for the workers' children, park
concessions to plan and organize, campgrounds and infrastructure to
build—all told, a decade-long array of tasks to create a park of this
magnitude. Kris, like Doug, also remained constantly on the prowl
for new land acquisitions, though her interest gravitated toward the
grasslands habitats that first entranced her. Pumalin would not be
their only big project in Patagonia. But it would remain Doug's personal favorite.

Much of Pumalin Park was intended to be remote and remain in
its original wild state. The closest city to the park's northern bound-
ary is Puerto Montt, 130 miles to the north, and a twelve-hour ferry
ride down the Corcovado Gulf on Chile's southern Pacific coast. The
Tompkinses' house is within the park. Nearby, at the ferry landing on
the southern shore of the fjord, they built an ecotourism complex and
welcoming center, Caleta, designed by Doug. A showcase for modern
park design, the wooden buildings were constructed from naturally
fallen trees and deadwood harvested from his lands. Local arts and
crafts were sold in the park store; organic honey was sold from the
organic beekeeping operation Tompkins set up inside Pumalin; a
restaurant featured foods grown at organic farms on the park's periphery. There were campgrounds, lodges, a schoolhouse for staffers'
children, an information center, and a series of trails into the southern section of the park and to the sea lion rookery. Caleta became the
model for the entire park infrastructure, putting parks elsewhere in
Chile—and in the United States, for that matter—to shame, a selling
point for local visitors and international tourists alike.

At first, none of this generated much controversy, attention, or
interest. For decades, Chile's philosophy on foreign investment had
been to welcome any and all, with few or no restrictions. Perhaps that
was where the later allegations of secrecy really came from—not from
any particular silence on Tompkins's part, but because no one really
took notice. Foreigners were buying up land in Chile all the time,

some of them with conservation in mind, too: George Soros, Luciano Benetton, Ted Turner. It was impossible to keep track. Tompkins, after all, was creating jobs and helping build tourism in the area, and more than 90 percent of his mountainous, densely forested land was too rugged for real development anyway.

But there were a few unwritten rules for foreign investors, and Tompkins broke a big one when he criticized and then sued a powerful Chilean salmon farmer in 1994. And that was when the "secret" of Tompkins's conservation efforts entered the public consciousness, and the inevitable conflict took root between his zero-development conservation ethic and the long-held Chilean dream of a million-strong force of settlers taming Patagonia, much as the old West had been tamed in America, marching down a Southern Highway that never seemed to get built. It was not a realistic dream, but its romantic appeal ran deep.

It started, as most of Tompkins's ideas and conflicts usually begin, out in nature. Tompkins liked to kayak the frigid waters of Reñihue for daily exercise and to explore his property's untamed vistas—different every day, and yet always the same, vital, quiet, alive in an elemental sense that he felt San Francisco and Millbrook could never be. One day while paddling his kayak he spotted a sea lion carcass, then another, their heads severed. Just about every day after that, he saw more, vile and unnatural. Soon the beaches in the area were littered with the bodies, along with trash, toxic waste, and dead fish, despoiling the land Tompkins was so painstakingly restoring. Furious, he began asking around and watching the salmon operation in the waters near Reñihue, and he soon determined the *salmoneros'* security guards, all heavily armed, were killing the animals en masse to keep them from feasting on the penned fish. They'd shoot the marine mammals in the skull, then behead them to avoid leaving ballistics evidence behind. This gruesome practice assumed someone would eventually investigate the killings, but that had never happened, at least before Tompkins began agitating. Chilean laws protected the endangered sea lion, he complained, but there was virtually no enforcement. Chile's scant environmental laws and its version of the United States' environmental protection agency were either advisory or unfunded and toothless. So Tompkins took matters in his own hands and offered a reward for anyone with information about the dumping and sea lion massacres. Tompkins had gone to war specifically with the Fiordos Blancos salmon company, which operated a sizable salmon farm right off the Reñihue coast, but he also complained publicly of the environmental devastation inflicted by all salmon farming operations, an open secret few cared to acknowledge.

However legitimate his claims might have been—and the Chilean courts soon awarded an injunction against Fiordos Blancos in an initial finding that Tompkins's case was sound—the public perception was different. What local residents and legislators saw was a rich foreigner bullying a homegrown industry that Chileans looked on with pride, and that supplied much-needed jobs in an otherwise economically depressed region. It was as if a foreign investor had bought up half of Nebraska, then started complaining that corn produced through industrial agriculture was bad for the environment and harmful to Americans' health. He might have valid arguments, but the locals would still want to see him run out of town on a rail.

Chile, of course, has no native Atlantic salmon—the fish is an exotic species anywhere below the equator, and in any case Chile is a Pacific Ocean country. But the vast coastal landscape of Patagonia, Chile's relatively toothless environmental regulations, and the country's welcoming policy toward just about any foreign investment made it a natural magnet for businesses interested in copying Norway's successful salmon farming industry. Chilean salmon farming was just beginning a period of phenomenal growth when Tompkins began his salvo. Within ten years, it would be Chile's number two export (behind only copper), a $1.5 billion business employing 17,000 Chileans and a crucial part of the economy—not to mention the main reason Wal-Mart can sell cheap salmon. Unfortunately, salmon farming, particularly as it has been practiced in Chile, is one of the most environmentally destructive forms of aquaculture ever conceived, packing many thousands of fish into crowded pens, doused with pervasive quantities of antibiotics that contaminate all sea life in the area, where the ocean floor becomes thickly coated with a miasma of uneaten fish meal and fecal waste. These salmon farms create vast dead zones in the ocean where nothing can live, and they produce salmon so anemic and unnatural that their pale flesh had to be artificially colored to make them marketable. Tompkins didn't limit his attack to the ecological harm the farms caused; he also openly argued that the salmon operations made no sense economically. Salmon farms consumed far more usable protein than they produced, by as much as a factor of four, according to some studies. It was the ultimate in unsustainable production, Tompkins argued. Such farms could exist only because the companies were allowed to operate with complete disregard for the costly environmental damage they wrought.

But Tompkins overplayed his hand: Being correct was not the same as winning. Salmon farming had created 100 times more jobs in Patagonia than his park-building programs could ever hope to generate. Political and popular support for the industry was huge. The argument for conservation, for deep ecology, could not match the reality of jobs and money in a country where 30 to 40 percent of the citizens were living on less than ninety dollars a month.

The general manager of the salmon farm, Patricio Quilhot, formerly a colonel in Pinochet's feared secret service agency DINA,[1] refused to settle the matter out of court, even on the straightforward issue of polluting and dumping on Tompkins's land. Instead Quilhot fought back, and he knew exactly where to land a punishing blow: He lashed out at Tompkins's land purchases and park plans, calling him a threat to national security.

In a country insecure about its borders, its precarious democracy, and its place in the world, there were few more serious allegations that could be leveled against a foreigner—and Quilhot's claim, made in

1994, was quickly echoed by his allies in the military, the government, industry, and the media. The official investigations of Tompkins that followed were inevitable: His picture became a fixture over ominous headlines in the daily papers, as legislators debated everything from withholding tax breaks from donors of parklands to expropriating Tompkins's lands outright.

"The country is divided into two and the guilty part is a North American who doesn't even live in this country," fumed *Qué Pasa*, a conservative weekly in the capital city, Santiago. "His objective is, to say the least, dark, covering a vast territory from mountains to sea."

And so Doug Tompkins and Chile began a decade-long war over a most unusual question: whether the multimillionaire would be allowed to create Chile's greatest national park and then give it away.

4.

image and reality

The wild rumors started circulating soon after the intelligence agent turned *salmonero* uttered the phrase "threat to national security." It was said that the crazy gringo planned a nuclear waste dump, a second Israel, an illegal gold mine, an Argentine military outpost. He believed in zero population growth, in one-world governments, in the idea that trees were more important than human lives. He was a socialist, a Zionist, a fascist, a CIA agent, a foreign infiltrator, an ecoterrorist. Tompkins found himself besieged by a bewildering and potent mixture of right-wingers, leftists, nationalists, the military, industrialists, xenophobes, neo-Nazis, the president's cabinet, and even the Catholic church, all incensed by the American interloper and his dangerous ideas. He was warned that his phone was tapped. He received death threats. Landowners previously receptive to his offers to buy them out became skittish. Military helicopters and jets buzzed his farmhouse in Reñihue, sometimes as low as 100 feet, rattling the windows, spooking the animals, scaring the boys and girls in the school the Tompkinses had built for their park employees' children. When Tompkins complained, Chile's minister of the interior snapped, "This is our national territory and airspace, not Tompkins's. The government is not harassing him; it is only making sure the laws are obeyed." There hadn't been such tumult over troublesome Americans arriving and buying property in Patagonia since 1901, when Robert Leroy Parker and Harry Alonzo Longabaugh—better known as Butch Cassidy and the Sundance Kid—settled down for a few years as gentlemen ranchers just across the Argentine border from Pumalin. Butch and Sundance, it was said, got a much warmer welcome when they came to town, at least until the Pinkerton men showed up.

Tompkins tried to joke about the sudden notoriety: "Hey, Kris, we have to get to work on that tunnel through the Andes," and the two of them would laugh. But it wore on him. Most of Chile's old-growth forests were gone. He now owned a substantial portion of what was left, including the single largest forest of *alerce* trees in the world. He was trying to save something important, and he had been sure that good intentions—the value and beauty of what he was trying to accomplish—would speak for themselves. Through the filter of his sense of image-making, trained at Esprit, Tompkins saw his idea as something that sold itself. "It is pretty hard for a country to turn down a gift of 300,000 hectares," he would say.

The problem was the isolation and lack of population that made his conservation project possible in the first place: Very few people had firsthand knowledge of what he was doing. Most Chileans literally could not see, touch, or feel the park in progress. That left only an increasingly dark, negative portrait of Tompkins painted by the press, priests, and politicians to fill the void. Soon the controversy jeopardized his application to have the government designate his land a tax-exempt nature sanctuary; without this designation he could not realize his dream of donating Pumalin as a national park.

Aware of his peril, he did what he knew best: He went to work on his image. He began holding town meetings in the towns nearest Pumalin—Puerto Montt and Chaiten—and speaking to environmental groups and student organizations at regional universities. He put together a slide show of his land and the work underway to build a world-class park, his painstaking restoration of damaged lands, the organic farms being run by and for Chileans—without tunnels, nuclear waste, or Israeli spies. Gradually, some of the locals were assuaged. College students grew particularly enamored of Tompkins and his conservation message, hanging on his words and his explanation of deep ecology, though there was a bittersweet aspect to their expressions of gratitude for his good works. One college senior commented, "What you are doing in Patagonia is so important. But I can't help but wish it had been a Chilean with this vision, rather than a foreigner. We should have been able to do this ourselves."

But town meetings could only go so far. Nationally, Tompkins faced a difficult battle. Chile had only just emerged from the long, tyrannical, violent rule of the dictator Augusto Pinochet, who had been supported for years by the United States' meddling and intrigue—part of the reason that the rich and influential Tompkins was greeted with suspicion and paranoia. This was compounded by Pumalin Park's location, which happened to be both strategic and vulnerable to attack. Chile is the longest, narrowest nation in the world, and so its defense is a tactical nightmare—Pinochet had spent much of his rule stoking paranoia about the threat of invasion. The country is 265 miles across at its widest point, and the narrowest section, in Region X,[1] is a mere sixty miles from the Pacific coast to the border of Argentina, which has long had rocky relations with Chile.[2] This geographic wasp waist also happens to lie within the heart of Doug Tompkins's land. The gringo with the tourist visa really did slice the country in two. "Imagine if I went to the U.S. as a tourist, bought land from the Atlantic Ocean to the Gulf of Mexico in Florida, and said you can't travel across it by road," observed one of Tompkins's most vociferous critics, the Chilean national senator Antonio Horvath, an engineer, a staunch friend to the salmon industry, and an advocate of aggressive development in Patagonia. "They would throw me out. Well, that's what Tompkins is doing."

Horvath's analogy struck a raw nerve in many Chileans, but not because it provides a compelling argument against conservation and for national security—even Horvath admits Pumalin is so rugged that no army could pass through it. Horvath's argument resonates because Chileans believe it highlights American hypocrisy. It's not hard to imagine the nationalistic outrage and calls for protective legislation if a foreign version of Tompkins suddenly appeared in the United States. Just picture an outspoken foreign critic of industry, the global economy, and conservative politics, who bought up immense and beautiful swaths of Colorado or Utah or Georgia—then said no to roads, cars, ATVs, hunting, and development, all in the name of saving the environment. How long would it take the U.S. Congress to start investigating? For politicians to start condemning? Why, then, Horvath demanded, should Chile behave any differently?

By the time controversy enveloped Tompkins, Pinochet was out as president, but one aspect of his rule remained gospel in Chile and was embraced across party lines: his pro-growth, free-market policies, which relentlessly emphasized development, exports, aggressive extraction of Chile's natural resources, and deregulation as the best path to prosperity. In 1995, the newly elected president, Eduardo Frei, made it clear those policies would continue in the newly democratic Chile. No environmental or conservation project, Frei promised, would get in the way of economic growth—and certainly no foreign environmentalist named Doug Tompkins, whose philosophy of local economies, less consumption, and less extraction was anathema to the Chilean outlook. The new president of Chile used the worst term he could to describe an American in their midst: He called Tompkins *arrogante*. Arrogant.

In the United States, Tompkins is revered in conservation and environmentalist circles; Henry Paulson, the former chairman of the Nature Conservancy and secretary of the treasury during President George W. Bush's final two years in office, would later say of Doug and Kris Tompkins: "I know of no two people anywhere that have so completely and totally dedicated their lives, their energy, their talent, their money to the cause of conservation. And they get important, tangible results." In Chile, the news media seemed to seek out only the farthest-right critics of environmentalism to comment on Tompkins. An activist from a conservative think tank, the Center for the Defense of Free Enterprise, told one Chilean newspaper that Tompkins's philosophy of deep ecology was "antihuman and pits human rights and progress against the rights of nature." He painted deep ecology as a radical, dangerous fringe idea that sees human life as essentially worthless, that wants mankind to return to the Stone Age, and that has even inspired ecoterrorism in the United States, elevating environmentalism to a kind of antihuman religion.

This dark view of deep ecology gained considerable traction in the battle against Tompkins, even though—or perhaps because—the criticism was dishonest in tone and substance. As a philosophical ideal, Tompkins's deep ecology "platform" is hardly the stuff of a terrorist screed. Its most significant pronouncements, though couched in terms of morality rather than science, are in line with the sorts of responses scientists say the battle against mass extinction and global warming demands, and are in keeping with current understanding of the world's ability (or inability) to sustain current levels of food, energy consumption, and population. The problem is that as a practical matter, this eight-point platform—written by the founder of deep ecology, Arne Naess, and the environmentalist George Sessions; and published in 1993 by Doug Tompkins in *Clearcut: The Tragedy of Industrial Forestry*—is deeply disturbing to those who would deny the existence of anything like an impending environmental disaster, or who wish only to maintain the status quo, or who, like the Catholic church, view criticism of human overpopulation as potentially "anti-life." And it happened that the long-planned *Clearcut*, a collection of stunning and horrifying photographs and stories of industrial harvesting of forests, was published just as Tompkins's image problems in Chile became acute. One Chilean archbishop read about the platform and called Tompkins an "anti-Christian pagan," and other prelates soon took up the call.

So what is this fearsome belief system? The deep ecology platform, as published in *Clearcut*, states:

1. The well-being and flourishing of human and nonhuman life on Earth have value in themselves (synonyms: inherent worth, intrinsic value, inherent value). These values are independent of the usefulness of the nonhuman world for human purposes.
2. Richness and diversity of life-forms contribute to the realization of these values and are also values in themselves.
3. Humans have no right to reduce this richness and diversity except to satisfy vital needs.
4. Present human interference with the nonhuman world is excessive, and the situation is rapidly worsening.
5. The flourishing of human life and cultures is compatible with a substantial decrease of the human population. The flourishing of nonhuman life requires such a decrease.
6. Policies must therefore be changed. The changes in policies affect basic economic, technological, and ideological structures. The resulting state of affairs will be deeply different from the present.
7. The ideological change is mainly that of appreciating life quality (dwelling in situations of inherent worth) rather than adhering to an increasingly higher standard of living. There will be a profound awareness of the difference between big and great.
8. Those who subscribe to the foregoing points have an

obligation directly or indirectly to participate in the attempt
to implement the necessary changes.

The third plank in the platform is the primary source of controversy, from which all the other ideas in the platform (and objections
to them) emanate: the notion that humans have no inherent, absolute
right to exploit and lay waste to nature. Philosophically it's hard to
argue against this proposition, because its opposite would hold that
humans ought to feel free to pollute, destroy, and drive species to
extinction even when such action is not vital, but simply more convenient or profitable. As a practical matter, however, this third plank flies
in the face of human history, religious teachings about man's earthly
dominion dating back to the Old Testament and beyond, and the way
we live today. The related notion that there are too many people in
the world (a scientifically uncontroversial position) generates similar expressions of horror; Tompkins's opponents falsely argued that
it meant deep ecologists advocate mass abortions, mandatory birth
control, forced sterilizations, and worse. And the idea that quality of
life need not be seen as a measure of how much money or property
a person has (which is how standard of living is currently measured)
baffled and alarmed many people, as did the statement that the future
needed to be "deeply different from the present." Critics said this
added up to something no one should accept: a Luddite view of the
world, a turning back of the clock, and an end to progress.

Caught off guard by the vehement opposition to a donated park
worth $35 million, Tompkins responded to his critics unevenly. The
"image director" at times compounded his image problems with
angry diatribes that tended to create more sympathy for his opponents
than for himself: "All we are trying to do is make a sizable national
park in an area that is largely untouched, and suddenly I find myself
embroiled in a tremendous debate over abortion. . . . It's outrageous.
The idea of having wild areas for their own sake is incomprehensible
here." Such outbursts were gleefully reported in the press. Pundits at

Chile's two leading conservative daily newspapers—*El Mercurio* and *La Tercera*—were only too happy to paint him as an American bully right out of central casting. Meanwhile, few defenders were willing to side with the foreigner who found Chileans incapable of comprehending his generosity. His picture on a front page was enough to increase newspaper sales, and the papers always seemed to choose an unflattering photo, a craggy face frowning beneath a halo of gray hair and a black beret.

But when he could suppress his inner CEO and his impatience with Chile's chaotic democratic process, Tompkins was able to reach out more effectively. He discussed his beliefs with mayors, city councilmen, provincial governors, business leaders—and even the salmon farmers. Week after week, he took groups of legislators and government officials up in his Cessna, the bush pilot showing off his lands, his forest restoration work, the organic farms owned by local residents who had swapped their old, worn-out plots for new, more productive farms ringing the new park. Tompkins supplied these residents with seed and assistance, and gave them work as park wardens as well, so they could offer directions to visitors in remote areas of Pumalin, and stay on the lookout for the constant problem of illegal logging of the valuable *alerce* trees. Tompkins would dip below the clouds and see someone working a field, and he'd abruptly bank and descend, the legislators gripping their chairs as he bumped to a stop on a grassy pasture so the legislators could talk to the farmers themselves. Yes, yes, it was true, the farmers would say, they were doing better now than they had ever done before. We worried at first about Tompkins and his park, but he has treated us fairly. The legislators knew that not everyone was so happy, so they would ask pointed questions, hoping for a chink in the armor, some hint that something was amiss, but the farmers would shrug, smile, and shake their heads. One said, "Look. We have food, a house, good land. My son is going to school now—he has a future, instead of fighting, fighting just to stay alive. Why do you want to find something wrong with that? For us, it is good."

What could the legislators do but nod and shake hands and pile back in the plane so Tompkins could swoop over another valley and point out the native plant nursery, part of his project "Alerce 3000"—in which he is planting a new forest's worth of the rare, long-lived *alerce* trees, row after row. They are so slow in their growth—just a few inches a year—that the project will begin to mature only in the year 3000, he told them. The Chileans didn't know what to make of this soft-spoken yet forceful man then in his late fifties, fit and confident at the plane's controls, part PR man, part philosopher, part tour guide, part mountain man, part millionaire, and part madman. Who laid plans for the year 3000? As legislators, they could barely agree on plans for the next fiscal year. Then they listened as Tompkins explained that all the fuss over deep ecology was nonsense: The point of the platform was that it is self-evidently wrong to destroy life and nature except when there is no other choice. It is foolhardy, he would say, to choose coal over solar, global over local, dirty over clean, when environmental and climate disasters are fast approaching. Any population scientist, geographer, nutrition expert, or relief agency in the world would agree that there are more people than the earth can sustain—out of the 6.7 billion, on any given day, one-third are living in poverty, one-fifth have no access to clean drinking water, 12 percent are starving to death. But, of course, Tompkins would soothe, we all know that if population is to be reduced, if society is to be transformed, it can come only over the course of many decades, perhaps centuries. "I want to raise the consciousness of the world," he said.

It all sounded so reasonable that the visitors invariably walked off the little plane impressed, less angry, less concerned about the wilder allegations against Tompkins—though not necessarily converted.

In March 1995, a day after his fifty-second birthday and despite efforts to reach out to his critics, Tompkins picked up the paper at his town home in Puerto Montt and read a breathless account of how conservative legislators from Valparaiso—the Patagonian city where he hoped to buy the last 30,000-acre piece of land to complete Pumalin Park—had returned from their fact-finding mission to denounce him in the halls of the congress as a dangerous interloper. Despite testimonials to the contrary, they claimed he had been pushing farmers and indigenous people off their lands, threatening them with lawsuits and property disputes if they refused to sell. Now Tompkins would be summoned to the Chilean senate to explain himself. The "republic" of Tompkins needed to be reined in.

Tompkins responded by presenting a slide show and question-and-answer session in Valparaiso in the days before his Senate appearance. Slide after slide, drawn from his book *Clearcut*, showed devastated forests and scarred lands throughout North America. Then the slide show switched to aerial views of ravaged Chilean forests: trees reduced to logs, logs reduced to enormous oceans of wood chips, the hills stripped bare, roads cut through forests like ugly wounds. The implication was that the threat came not from Tompkins but from the politicians and companies that wanted to do to South America what had already been done in the north. The slides from *Clearcut* were contrasted with images of the startlingly blue lakes and fjords of Reñihue, the *alerces* of Pumalin, the dense green landscapes. "This is special," he told the gathering. "It is one of the last virgin reservoirs of the world and I intend to see it preserved." He promised he would soon publish a Chilean version of *Clearcut*. At the same time, he underwrote several other environmental book projects in Chile while doling out more than $30 million in grants to Chilean environmental organizations—raising their profile and clout, and establishing a viable conservation lobby for the first time in Chilean history. "The truth will be told," he promised.

"Douglas's biggest problem is that he has become part of a soap opera, fed by envy, one of our great national pastimes," the botanist Adriana Hoffman told the press—one of the few voices in Chile speaking out for him back then, when it was a risky thing to do. "Doug Tompkins is doing the country and the planet a favor. . . . Thank God someone is exercising stewardship to preserve a little bit of what is left of the world's frontier forest."

Tompkins dutifully traveled to Santiago and met with legislators on the left and right, government ministers and bureaucrats, and presidential aides, once again explaining his land purchases, his philosophy, and his goals. He produced documents to show that he had paid small landowners market value or more for the land, even when the residents lacked clear title, and that most of his land had been purchased from large companies and foreign investors. Early on, Tompkins had fired several employees for bullying some of the squatters, and he had taken his boat out and visited them personally, apologizing and offering to pay for real estate attorneys to get their land titles in order, so that they could choose to sell or stay as they saw fit. The current crop of about a dozen small landowners who had recently complained to the legislators were actually squatters on land next to Tompkins's holdings, illegally occupying small farms spread across 74,000 acres owned by the Catholic University of Valparaiso. The university had been trying for years to evict the squatters, and although it was true that Tompkins was negotiating to buy the land from the school, he said he had nothing to do with the attempted evictions.

Instead of placating his most ardent opponents, this explanation both incensed them and provided a new line of attack. Now the right-wing Independent Democratic Union Party demanded that the Chilean government petition the university and, if necessary, the Vatican, to prevent Tompkins from buying the land owned by the Catholic University. The party wanted a full investigation of the deep ecology movement and its promotion of a "world of less people." They had heard stories, the legislators reported, of couples working on Tompkins's properties who had been asked not to have children. Was the school in Caleta preaching zero population growth and the idea that human life was no better than animal life? The left-leaning Christian Democratic Party soon joined the call, and a congressional committee was appointed to conduct yet another investigation and visit to the Pumalin area, to see if Tompkins was promoting the "radical" idea that women had complete control of their bodies, including a right to birth control or abortions. "All this is totally incompatible with Chilean legislation," one legislator, Sergio Elgueta, thundered. They neglected to mention that the school at Pumalin was supervised by inspectors from the Chilean education ministry, who had approved the curriculum as conforming to national standards.

The legislators also expressed grave concerns about the fate of 150 workers whose livelihoods depended on the Fiordos Blancos salmon farm operating in the waters of Reñihue Fjord. Workers there complained that Tompkins had closed an access road through his property, leaving them no direct commute to work. Worse still, the farm operators threatened to shut down entirely if Tompkins won the court injunction he was seeking against them—everyone would be laid off. The implication was that Tompkins was being unreasonable and uncaring, and the legislators suggested that the appropriate solution to such outrages would be to regulate—perhaps ban—large land purchases by foreigners when important national interests were at stake. When they introduced legislation to accomplish this goal, the regulations they proposed would have posed no obstacles to foreigners who wanted to buy large tracts for logging or mining; only land purchases for conservation would be closely regulated under their proposal.

Again, Tompkins appeared to counter the attacks. A road through the park had been closed; that was true, Tompkins said. However, he wasn't trying to interfere with workers' commutes; he was blocking salmon farm workers from disposing of waste, including beheaded sea lions, at an illegal dump they had created without permission on the Reñihue property. That was one of the reasons a provincial judge had issued a preliminary injunction against the salmon farm— to protect Tompkins and his land. Tompkins also pointed out that

the manager who had first raised the concern about Pumalin Park
and national security had been thoroughly discredited. He had been
named in court filings as a member of one of Pinochet's death squads
and was alleged to have been an accomplice in the kidnapping and
murder of a United Nations diplomat. Finally, Tompkins pointed to
evidence that the legislators on the right and the left who were most
vociferous in their attacks on the park, who had raised the allegations
of promoting abortion and of bullying landowners, and who wanted
to legislate against foreign eco barons, were part owners of the same
salmon farm that had been dumping on Tompkins's land and slaugh-
tering sea lions.

With those revelations, the very public attacks on Tompkins in
the Chilean congress suddenly became too "sensitive" to be discussed
in public, and were passed on to a presidential blue-ribbon commission for further study. Tompkins, meanwhile, strengthened his hand,
creating a Chilean foundation, run by Chileans, to take ownership of
Pumalin Park. He persuaded the respected bishop of Ancud in Patagonia, Monsignor Juan Luis Ysern, to become president of the new
foundation's board, in order to blunt the notion that Tompkins was
promoting an anti-life, anti-Catholic message. Ysern said he had been
leery of Tompkins, his project, and his philosophy, but had come to
"wholeheartedly support" the park plan and conservation efforts. He
dismissed the allegations about Tompkins as propaganda and preju-
dice. "It doesn't happen every day that a rich man should invest without seeking his own personal wealth," Ysern said. Tompkins scored
another victory when he won his court case against Fiordos Blancos,
whose owners promptly sold out to a large Canadian-owned company
with a far better environmental record.

But just as the blue-ribbon commission prepared to dismiss the
allegations against Tompkins and recommend that Pumalin receive
status as a sanctuary, President Frei bowed to pressure from Chile's
powerful military and right-wing politicians by blocking the sale of
the 74,000-acre tract owned by the Catholic University of Valparaiso,

long prized by Tompkins. The university had been prepared to accept Tompkins's offer of $1.7 million for the land. The parcel, known as Rancho Huinay, consisted of virginal rain forest that had never been logged, except for a few cuts and burns by the small band of squatters, and Tompkins needed it to complete his master plan for Pumalin, as it provided the missing link joining the southern and northern ends of the vast park. Now, the president decided, the government of Chile would buy Rancho Huinay instead, and then determine what balance of conservation and development would be appropriate. No sanctuary status would be granted to Tompkins. And waiting in the wings was Chile's most powerful energy company, Endesa, which had a different vision for Patagonia, one of massive dams, hydroelectric stations, and thousands of miles of transmission towers carrying power generated in the wilds to the urban centers in the north. Chile's rapid growth had led to an energy crisis, and the head of Endesa, José Yuraszek, had the president's ear. He wanted a "Yuraszek Park" in place of Pumalin, where he could bulldoze and develop at will. Tompkins's foes were ecstatic. Instead of Tompkins slicing Chile in two, the opposition had sliced his park in two—the north and south halves of Pumalin would never be contiguous. "I have been beaten up by the government for wanting to invest millions of dollars in a project the country really needs," a disheartened Tompkins complained at a press conference in July 1995. He might just close off the lands and keep them private, he vowed, if that was what it took to save the last wild places in Patagonia.

But if Tompkins's harshest critics were satisfied, others decried President Frei's decision to overrule his own blue-ribbon commission. Ricardo Lagos, who was the minister of public works and a leading candidate to succeed Frei, railed against a government that had no problem welcoming foreign investors who exploited Chile's resources, yet mistreated the only one who wanted to preserve valuable resources. Major media in the United States began to take notice, then—*The New York Times*, the *Washington Post*, and others[3] brought news of the Tompkinses' conservation efforts and the controversy to an American audience wider than the environmental community that had long cheered them on. Soon the U.S. ambassador to Chile and the House minority leader Dick Gephardt intervened, making it clear that pending trade talks between the two countries could be jeopardized if it seemed the rights of Americans in Chile were being trampled. Suddenly the "threat" that Tompkins and his deep ecology posed to Chile seemed hardly worth talking about; President Frei put his plan to buy the Huinay property on hold and reopened negotiations with Tompkins on the fate of Pumalin, eventually promising, in 1997, to introduce legislation to make Pumalin a protected sanctuary.

Four more years passed, and the promised legislation did not see the light of day. Although the government had backed off, the university refused to sell Huinay to Tompkins and in 2001 instead sold it to the energy company Endesa. Endesa vowed to preserve the land's natural beauty, but Tompkins and his Chilean allies saw it as a beachhead in a war over the future of Patagonia's rain forest. Tompkins called a press conference to say he finally was pulling the plug on donating Pumalin Park to Chile. He would keep the land private and turn his attention and money elsewhere.

"I am fed up with waiting," he told the newspaper *La Tercera* in August 2001. "I have better things to do with my life."

In truth, Doug and Kris Tompkins had not been waiting at all—they had simply shifted their attention to other conservation efforts designed to preserve equally important wildlands, but in locations less likely to provoke nationalistic emotions and protests. For better or worse, the landscape of Pumalin occupied mental territory that, to Chileans, was spiritually akin to Americans' perceptions of California a century ago. Tompkins had bought up a big piece of Chile's "California dream"— its "Patagonia dream." That he was trying to save it didn't matter. What mattered was that it was *their* dream. But there were plenty of

other places in Chile—and in other, lesser-known areas of Patagonia—
that presented golden opportunities for conservationists.

So, by the time a newspaper interviewer was hearing how weary
the Tompkinses had grown of the endless fight over Pumalin, Doug
and Kris were well into a spree of buying land and creating parks that
was unprecedented in the conservation history of any country.

First, with Doug's former trampoline instructor and Esprit partner, Peter Buckley, they purchased 208,000 acres of coastal cypress
forest south of Pumalin, and created two spectacular wildlife and
nature preserves: Corcovado and Tic Toc. Shortly before the purchase, Tompkins had flown Buckley up and down the Patagonian
coast, pointing at one magnificent vista after another. "That's for sale.
That's for sale," Tompkins would shout, veering the Cessna toward a
snowcapped volcano. "And that one with the glacier and river, that's
for sale, too." Buckley called Tompkins his "crazed real estate agent,"
but one gorgeous, green slice of forest and blue water caught his eye
for a homestead, and he was happy to buy up the much larger land
around it to donate. "You couldn't say no to Doug," he says. "And I
have a hell of a lot more money than I'll ever need. Why not do something good with it, something that lasts?"

There was no controversy in this region: Unlike Pumalin, the land
at Corcovado had been unoccupied since the 1930s, when the cypress
forests had been severely overcut to make vineyard stakes for the wine
country near Santiago. The land and climate had been too hard, and
the few settlers were long gone. Sixty years later, the second-growth
cypress trees were coming in strong, and what remained of the original forest still represented some of the largest stands of old-growth
cypress left in Chile. This land was quietly awarded the status of
nature sanctuary that still eluded Pumalin, and Tompkins floated the
suggestion that, combined with 500,000 neighboring acres controlled
by the Chilean military, it would make an amazing national park.
The government seemed to agree, and began laying plans for just

such a park, as if the enmity and conflict still going on at Pumalin did not exist.

Through their Conservation Land Trust foundation, the Tompkinses next purchased 6,000 acres of evergreen forest in the Melimoyu Preserve on Isla Magadelena, south of Corcovado and adjacent to a seldom visited, poorly maintained existing national park that they hoped to improve. Another 65,000 acres of virgin island forest off the coast of Tierra del Fuego were purchased to form the Cabo Leon Preserve, which the Tompkinses donated to a private Chilean conservation group they supported, the Yendegaia Foundation, run by their old friend Adriana Hoffman. Another 100,000 acres, along the old disputed waters of the Beagle Channel, were purchased with Buckley and a few other investors when a large parcel of coastal mountain forest went on the market as the owner, a convicted drug dealer, desperately tried to pay his legal bills. This Yendegaia preserve, also entrusted to Hoffman's foundation, is home to dozens of endangered plant and animal species, and forms an important land bridge between two existing national parks—part of Tompkins's strategy to ease the extinction crisis in Patagonia by establishing wildlife corridors that span the region. All the properties represented unique landscapes and habitats, and all had been in danger of being mined or logged into oblivion if left unprotected.

The Tompkinses next added to their portfolio three huge parks and preserves in Argentina. The first was the 450,000-acre Estero del Ibera in Corrientes, Argentina, part of the cloud forest in the northwest of the country. Much of the land they ended up preserving in Corrientes was to have been cleared and remade into a giant industrial tree farm of nonnative pines, a practice that Tompkins loathed, because planting nonnative species frequently introduced new parasites, insects, and diseases, ultimately damaging habitats and hastening extinctions throughout whole regions. Tompkins had bought the land out from under the timber company at the Argentine government's invitation, a very different experience for him and a relief from the thorny relations with Chile's leaders. The nature preserve of wetlands and steppe that the Tompkinses established there is home to the endangered maned wolf, the capybara (the world's largest rodent—a 140-pound cousin to the guinea pig), the endangered pampas deer, the giant anteater, the crowned eagle, the tapir, and more than 360 species of birds.

To the south, the 37,000-acre El Rincon preserve of glacier-fed streams and badly overgrazed sheep and cattle lands is being gradually restored by the Tompkinses. They sold off the livestock herds after purchasing this Andean foothill property, in the shadow of one of Patagonia's great peaks, San Lorenzo. With the grasslands restored, biologists believe the endangered huemul deer will repopulate the area, spreading from an adjacent national park where they have been making a comeback in recent years.

As they accumulated land and conservation projects, the Tompkinses built a staff of nearly 100 laborers, biologists, veterinarians, farmers, lawyers, rangers, administrators, and others, supplemented by crews of student volunteers and ecotourists from the United States, who sign on for working vacations at the Tompkinses' properties, restoring ranch property back to its original, unfenced wild state.

The jewel of the Tompkinses' purchases in Argentina is the 155,000-acre Monte Leon, the nation's first coastal national park. Kris Tompkins took the lead on this project, purchasing one of Patagonia's most historic sheep ranches, selling off the livestock, establishing a wildlife preserve, then donating it to the Argentine national park system—with no objections from left, right, or center. Monte Leon, with its thirty-one miles of beachfront, encompasses one of the last completely wild coastlines outside Antarctica. It provides a home to more than 65,000 mating pairs of Magellanic penguins and important nesting grounds for giant cormorants, and its inland steppe is home to large populations of pumas, rheas, foxes, and the wild llama species known as the guanaco.

Kris Tompkins's personal favorite of their many projects—her version of Doug's Pumalin, her "opportunity of a lifetime"—came in the form of a 173,000-acre, badly overgrazed Chilean sheep ranch near the Argentine border, Estancia Valle Chacabuco. Well north of Reñihue and once the largest working ranch in Chile, Chacabuco is a diverse and biologically rich landscape, dominated by the grasslands and steppe that Kris loved from the moment she first visited Patagonia—a type of Chilean landscape unrepresented in existing parks and preserves. Acquiring Chacabuco had been the number one priority of Chilean national park officials for thirty years, but either the owners wouldn't sell or the cash-strapped park service couldn't afford to buy. The Tompkinses had a good relationship with the Chilean parks officials, but given the charged political climate and the Pumalin controversies, they rarely could work openly together. But Chacabuco would be different. Kris and Doug, with additional financing from Yvon Chouinard, beat out a "Stop Tompkins" business consortium made up of Chilean timber and energy companies, including the Tompkinses' old adversary, Endesa. Their last-minute bid for $10 million won the day, even as the anti-Tompkins forces were gloating in the newspapers that same morning about winning the property from its Belgian owner so they could develop rather than preserve it.

Through her separate Chilean-based foundation, Conservación Patagonica, Kris Tompkins launched a nine-year plan to restore the worn-out land. She sold off the sheep and assembled an army of volunteers from her old company, Patagonia, as well as crews of ecotourists who began coming to Chacabuco to live in rustic cabins and work hard, paying handsomely for the privilege of ripping out 500 miles of fences so that wildlife can roam free. The Chilean government, again in contrast to the acrimonious dealings over Pumalin, has agreed to merge existing military-controlled lands with the Chacabuco ranch property, which the Tompkinses will donate once their restoration work is complete. The final result will be a 650,000-acre Patagonia National Park. This region has a long history of tourism along with ranching, and so the idea of a park was far more welcome there than in Pumalin's forests—and the fact that Kris Tompkins, rather than Doug, was the point person for the park undoubtedly helped as well.

The new park will be of a size and scope that will rival America's Grand Tetons National Park. Every biome and habitat in Chile is present—grasslands, foothills, mountains, desert, steppe, riparian areas, lakes, rivers, wetlands, and old-growth forests. Most of the original species are still present, including the huemul deer, though many such species must be nurtured back from the brink of extinction through careful tracking, assisted breeding programs, and rewilding. Biologists from Chilean, European, and American universities have come to the area to study the endangered huemul; others are attempting to tag and track pumas to see how they behave without sheep to prey on. The region has been sheep-farmed for more than a century, so the herds' absence offers a unique opportunity to study how nature will respond. Kris Tompkins is leading a committee charged with developing a master plan for the park. There are a million things left to do, she says, thousands of invasive nonnative plant species must be removed, years of work lie ahead—and she sounds like a kid on her way to Disneyland every time the subject of Patagonia National Park comes up. "Just another two hundred miles of fence to go," she said brightly after returning from a horseback ride across the ranch.

"There's nothing I've ever done that's more important," she says. "We are protecting the things we love."

By 2005, Doug and Kris Tompkins had acquired more than 2 million acres for conservation in Chile and Argentina, and they continued to hunt for more—and to battle over land already in hand.

Pumalin Park remains the centerpiece, the most beautiful, the most remote, and the most troublesome. Finally, in 2005, with a new president in place—Tompkins's former supporter, Richard Lagos—and a new environmental chief, Adrianna Hoffman, Pumalin was granted the long-promised status of nature sanctuary. The park by then was open to the public, and on track for eventual donation to the people of Chile, or to become a permanent land trust. The controversy and distrust faded and Tompkins's stock rose in Chilean public opinion polls as visitors poured into his parks.

The relief was short-lived, however. A few months after Pumalin became a sanctuary, a consortium of power companies led by Endesa announced a huge new project—four enormous dams on Patagonia's two largest and most wild rivers, the Pascua and the Baker, as part of a large-scale hydroelectric project to provide power for development to the north. If the project is carried out as proposed, thousands of acres of virgin forestland will be flooded, and even more will have to be cut down for service roads, substations, and power transmission lines.

There were several routes that the power lines and roads could take, some of them far less destructive than others. The proposal Endesa favored, perhaps unsurprisingly, runs right through the heart of Pumalin Park.

"I'm in the fight of my life," Tompkins said in 2008, and it's hard to tell, as he speaks, whether he is angry or excited at the prospect.

# Part Two

# of hot spots and hooters

Kieran Suckling, Peter Galvin, and the Center for Biological Diversity Reinvent Environmentalism

Nature hates calculators.

—RALPH WALDO EMERSON

thinking like aldo

Between the asphalt sprawl of the Los Angeles basin and the fertile flatlands of the Great Central Valley, where the Okies flocked during the Dust Bowl days and the *Grapes of Wrath* was set (and banned), a vast and surprising wilderness still thrives. Broad pastures, granite-studded hillsides, and icy-blue mountain lake water all lie within an hour's drive of L.A. smog and concrete, hidden in plain sight as traffic snakes by on the one major freeway through the weathered grandeur of the Tehachapi Mountains. Ancient oak groves offer cool shade, thick stands of piñon pines beckon in an incessant breeze, forests of twisted Joshua trees grope toward the clouds in gnarled supplication. In spring, the hillsides are carpeted in wildflowers, a gaudy sunset of color and fragrance: mariposa tulips, California poppies, baby blue eyes, Indian paintbrush, and owls' clover offer a last refuge for the vanishing wild honeybee so essential to the orchards and farmlands in the distant valley below. The landscape has barely changed for thousands of years, which is why more than eighty rare and endangered species still prey, roam, roost, flower, and raise their young here. Even the nearly vanished California condor still comes to forage in small, fragile numbers, where long ago whole colonies nested, the skies black with the colossal birds, their broad, dark wings beating the air with the sound of feathery thunder. Now wildlife biologists must help feed and care for North America's largest bird, because the condors cannot survive on their own. This last Southern California wilderness, this enormous blank spot on the map twelve times the size of Manhattan, provides an important barrier against extinction. There is no other place like it in California, and few to rival it on earth: To stand on a windswept hill at Tejon Ranch is to be at once humbled, enthralled, and saddened by vistas that in years past defined California and the West by their plenty, rather than their dearth. The United Nations has recognized the region as a vanishing geographic breed, one of twenty-five irreplaceable hot spots of biodiversity in the world—a designation reserved for just 2.4 percent of the earth's surface.

But designation does not necessarily bring preservation. Tejon Ranch also happens to be the biggest piece of privately held property in all of California, and its owners and investors have decided to build and build big.

Where cattle grazing and hunting have been the main activities for the past 150 years, the Tejon Ranch Company and its investors want to construct the single largest master-planned community California has ever seen—which is saying a lot for the state that raised scrape-and-sprawl construction to an art form. They intend to raise out of the wilderness an instant city of 30,000 homes and 70,000 people called Centennial, along with industrial parks, cargo terminals, shopping centers, and a separate resort and luxury home complex, Tejon Mountain Village, abutting the condor's historic nesting grounds—a supposedly self-contained and self-reliant community. Thirty thousand jobs could be added to a struggling local economy. Completing the project would demand, on average, the construction of one new house every eight hours, 365 days a year, for twenty years.

Sums of money as vast as the landscape are at stake—the raw land alone was valued at $1.5 billion in 1999, and if fully developed as currently envisioned, the ranch would be worth ten, twenty, or thirty times that amount, perhaps much more when all is said and done. To those who see progress in a bulldozer's blade and beauty in the taming of wilderness, Tejon Ranch is an irresistible plum, more than 250,000 contiguous acres lying a mere seventy miles from downtown Los Angeles, a straight shot up Interstate 5—the Golden State Freeway, as it's called, transformed into the ultimate driveway to the ultimate bedroom community. *The New York Times* admiringly described this one-of-a-kind plan for a one-of-a-kind landscape as "Playing SimCity for Real."

Standing in the path of this future Tejon Ranch is a relatively little-known environmental group with a small budget and outsize ambitions, the Center for Biological Diversity.

On paper, this scruffy outfit with the tree frog logo and the borrowed Tucson gem shop for a headquarters shouldn't have a prayer against the nearly limitless political, economic, and legal resources behind SimCity, except for the fact that during the past twenty years the Center for Biological Diversity has won close to 90 percent of its 500 cases. This unprecedented success rate has quietly transformed the American landscape, safeguarding hundreds of species from extinction and preserving millions of acres of wilderness. The center has taken down off-roaders and off-shore oil drillers, developers, and Detroit automakers, wolf haters and condor killers, and an entire alphabet soup of government agencies from Washington state to Washington, D.C., and as far away as Okinawa. The Center for Biological Diversity has fashioned itself into the most effective environmental operation you've never heard of, routinely outperforming the better-known and more moneyed conservation organizations in exposing corruption and official lawbreaking, then bending local governments, multinational corporations, and even presidents to its leaders' will. Even its most ardent detractors concede this is not hyperbole: It was the Center for Biological Diversity that finally forced the administration of President George W. Bush to concede, after six years of resolute denial, that there really is such a thing as global warming and that it is killing (among other species) the polar bear.

And yet, unlike the developers of Tejon Ranch, the center has never been the subject of an admiring profile in *The New York Times*. More typical is the *Wall Street Journal*'s twelve-hundred-word feature gleefully headlined, "Rancher Turns the Tables," which gave short shrift to the organization's remarkable string of victories and instead lionized the one person in twenty years who successfully sued them. The *New Yorker* magazine, in a 1999 piece by the usually astute Nicholas Lemann, portrayed the center's leaders and staff as humanity-despising destroyers of the great hunting, ranching, and cowboying traditions of the American West, closing the piece with this dark pronouncement:

> *They're outlaws. Outlaws cause trouble, alter the established order, and make authority figures angry. And, in the end, they get dealt with.*

This surprising metaphor seems especially ominous, given that the center's executive director has received repeated death threats, and it is curiously off base factually, too, because the signature position of the Center for Biological Diversity is not that laws ought to be disobeyed, but that the nation's existing environmental laws ought to be *enforced*. The group arguably can be called inflexible, infuriating, and litigious, and its leaders are at times—at least from their detractors' points of view—uncompromising, even extreme. Yet this idea the center pushes—actually enforcing laws rather than merely paying them lip service as whole species and ecosystems expire—is the precise opposite of the "outlaw" stance. It does, however, turn out to be a deeply unpopular position among the genuine environmental outlaws in government and industry, as well as the media and even some mainstream environmental organizations, whose corporate donors prefer gentler compromises with polluters and developers. But the center's staffers are used to being cast in the black-sheep and underdog roles, this eclectic hodgepodge of fifty-five underpaid and overworked lawyers, biologists, activists, and otherwise pissed-off ordinary folks, accepting the wrath of loggers, builders, ranchers, cowboys, developers, conservatives, journalists, and bureaucrats. "Oh, guess what, the *New Yorker* says we're outlaws who ought to be killed," the executive director informed his staff after the Lemann article appeared. Then, only half joking, he said, "Let's use that in our next fund-raiser."

Though few outside the rarefied world of environmental litigation are familiar with the center, evidence of its work is ubiquitous. When there is a news report about some nearly extinct bird or bear or sea turtle receiving official protection, or a habitat or forest preserved from development, or perhaps some new court ruling aimed at cutting greenhouse gas emissions or improving fuel efficiency for sport utility vehicles, more often than not, it's the result of a lawsuit, real or threatened, by the Center for Biological Diversity. Almost every species grudgingly listed by the Bush administration as imperiled since the year 2001—a total of eighty-seven—has been protected because the center used the courts to force the issue on a recalcitrant White House. Since the Endangered Species Act was passed in 1973,[1] 70 percent of the plant and animal species given protection under the law were listed because of efforts by the center. In the process, more than 70 million acres of wildlands have been preserved as habitats for these endangered species—an area nearly twice the size of all the national parks in the contiguous forty-eight states combined. The center's work has transformed overgrazed, trampled, and befouled federal lands that had been all but given to cattle barons and left for dead into thick, lush riparian forests. Thousands of miles of ocean waters nearly stripped of life have been made off-limits to deadly dragnets, bringing endangered sea turtles and depleted wild fisheries back from the brink of extinction. One court case pursued by the center halted logging not in a single habitat, not in a single forest, but in *every* national forest in the Southwest—all eleven of them—after it was shown that the feds were routinely breaking the law by giving loggers carte blanche to cut down ancient trees in environmentally sensitive public lands.

It is no exaggeration to say that the modern American environmental movement has been reinvented by the center, and especially by two of its founders and leaders—Peter Galvin, a former U.S. Fish and Wildlife Service owl expert, scientist, and mystic who devises intricately ruthless environmental campaigns; and Kieran Suckling, a former engineering student turned philosopher and amateur chef who combines an encyclopedic knowledge of animal species facts with a political opposition researcher's instinct for the other side's jugular. ("Boxing's not about hitting hard," Suckling likes to say, explaining the center's technique of burying officials with flurries of lawsuits, investigations, petitions, and press releases. "It's about jabbing the other guy over and over and over, before he has a chance to recover.") This amalgam of grit, talent, and willingness to repeatedly poke the entrenched and powerful has allowed these two iconoclast outdoorsmen to enter the ranks of America's eco barons—not because they have money, but because they don't seem to care about it, throwing themselves headlong into David-and-Goliath battles others have written off as hopeless. Their opponents (and, sometimes, other environmental groups that disapprove of the center's tactics) have never forgiven them for what usually happens next: They win.

Their latest target is the most threatening source of environmental damage, extinctions, and habitat loss yet—global warming—and Tejon Ranch is ground zero. The center's leaders envision curtailing or halting outright the planned development at Tejon Ranch as a milestone blow against climate change—an attempt to stop America's insatiable, unsustainable drive to pave over the last bits of her wilderness, which absorb greenhouse gases, and to replace them with new cities, carports, and commutes that pump out those deadly gases by the ton. The center wants to force developers at Tejon—and elsewhere—to quantify their contribution to global warming, and then do everything feasible to eliminate that impact, from installing solar roofs to creating electric-powered bus lines to mandating zero-emission vehicles for all residents. And if the developers refuse such "mitigation," the center argues, the entire project should be scuttled.

Of course, no developer has ever had to do such a thing. It's ludicrous, the Tejon Ranch people say. It'll never happen.

Except—with the Center for Biological Diversity, there's always an "except"—what if it turned out that there have been obscure laws on the books since the 1970s that *already require* builders to limit greenhouse gases and slow global warming? What if the only problem is that no one thought to ask?

Until now.

Before he created Pumalin Park, before he threw himself into saving Patagonia's wild places, Doug Tompkins built the Foundation for Deep Ecology so that he could provide seed money—sometimes a few thousand dollars, sometimes a few million, usually somewhere in between—to support activists who were trying to advance the ideals of deep ecology. He wanted to dispense grants to groups who were protecting biodiversity and "wildness," who were pioneering ecologically sound agriculture, and who wanted to combat the forces of globalization and what Tompkins calls "megatechnology." That meant the foundation did a little bit of everything that Tompkins cared passionately about. It bought a six-month blitz of weekly full-page ads in *The New York Times* as part of Tompkins's "Turning Point Project," decrying such issues as genetically engineered produce, the extinction crisis, oil spills, and "welfare ranching." The foundation organized and funded mass protests against the World Trade Organization (WTO), most famously in 1999 in Seattle, where 40,000 protesters forced the WTO to end a critical ministerial meeting in failure. It published evocative large format, photo-essay books such as *Clearcut* and *Fatal Harvest*. Most of all, the foundation helped pay the bills for some scrappy, hard-core, visionary environmental organizations, the sort of groups that corporate foundations often wouldn't touch, because they were perceived as too radical, extreme, or uncompromising. These groups were outlaws only in the sense that the Center for Biological Diversity was an outlaw: they didn't hesitate to go to court to enforce laws, irrespective of the politics of the situation or the popularity of the projects and programs that might be scuttled.

Tompkins sent millions to such groups as the antiwhaling Sea Shepherd Conservancy, the Rainforest Action Network, the Tides Foundation, the Earth Island Institute, Forest Guardians, and Friends of the Earth, and to more than 500 other environmental groups. Tompkins's foundation hosted a meeting of prominent environmentalists in 1991 who launched a new cooperative initiative they called the Wildlands Project. Tompkins provided more than $500,000 to kick-start this little-known, immensely ambitious project, a 100-year plan that seeks to knit together the immense and mostly unoccupied public lands in the western United States into a network of wilderness zones. Natural areas would be joined by wildlife corridors and buffers from civilization, in the hope that the Wildlands Project could eventually "re-wild" large portions of North America—a much larger version of the green corridor Tompkins helped create in Argentina at El Pinalito Park. Tejon Ranch, for instance, is seen as a potential (and critical) link in California's leg of the project. Michael Soule (who pioneered the discipline of conservation biology) and Dave Foreman (cofounder of Earth First!) first conceived of the ideas behind the Wildlands Project. Dozens of conservation groups, state agencies, and property owners have agreed to participate. But Earth First!—with its motto,

"No compromise in defense of Mother Earth," and its long history of protests, civil disobedience, and "monkey-wrenching" developments and logging that the group considers environmentally unsound—has long been a target of criticism, particularly from the political right, and Foreman's involvement raised instant suspicions in some quarters about the Wildlands Project. The project's re-wilding proposals are scientifically uncontroversial—the focus is primarily on restoring, connecting, and providing buffer zones for wilderness on public lands that have no permanent human residents. But critics feared a power and land grab, portraying the Wildlands Project as an attempt to trample on property rights, to limit hunting and off-road recreation, and to return half of North America to a depopulated, prehistoric condition. Web sites such as "The Wildlands Project Revealed," some subsidized by conservative foundations intent on balancing Tompkins's foundation, went up to denounce the project; the popular Free Republic site likened Wildlands Project organizers to a "green al-Qaeda." The ambitious project slowed considerably during the Bush administration, which agreed with the critics.

Tompkins gave millions more in grant money to media watchdog groups (including the Public Media Center and AdBusters); to his old mentor, the former adman Jerry Mander, who had founded the International Forum on Globalization to highlight concerns about the effect of globalization on democracy and public accountability; to food safety and anti-biotechnology organizations led by the attorney and author Andrew Kimbrell (a leader in calling for reforms in the beef industry and for safeguards against meat tainted by mad cow disease); and to the provocative economist Jeremy Rifkin, a critic of bioengineering and other so-called "Frankenstein" technologies, as well as an advocate for a new "hydrogen economy" to displace oil, coal, and natural gas.

Just about every group and activist the foundation funded was, in effect, a professional thorn in the side of one or another industry or government agency, or of Wall Street's conventional wisdom about economics and regulation; most of the recipients would have decried the global practices of Esprit, the source of the largesse distributed by the Foundation for Deep Ecology. Tompkins wasn't interested in giving money to organizations with artful brochures or with a reputation for playing nice with the other side; he wanted fighters.

His foundation found a most likely match in the Center for Biological Diversity: led by two guys with beards and hiking boots, who were as comfortable sleeping on a mountaintop ledge as Tompkins, who felt best when they were in wild places, who were convinced of their own righteousness and happy to fight for it—so long as it meant preserving a threatened piece of the natural world. Over the years, the foundation has given the center more than $200,000—up to $65,000 in one year. In the early days, when the center was paying its people less than the minimum wage and lawyers were sleeping on the office floor, that money kept the organization alive. The center is not quite so lean these days—it receives larger grants from several sources, and thousands of smaller donations through membership drives—but Tompkins and his foundation played an important role early on in sustaining what would turn out to be some of America's most effective environmental activists.

The story of the Center for Biological Diversity begins not with lawsuits or protests, but with the hooters.

In 1989, the U.S. Forest Service began sending owl survey crews into the Gila National Forest in New Mexico, one of America's most rugged and remote public forests, a mix of mountain, chaparral, meadow, and woods where the legendary conservationist Aldo Leopold wrote of learning to "think like a mountain." The government surveys were in response to the concern that heavy logging in the Gila was driving the gentle, brown Mexican spotted owl into extinction— a concern first raised by Robin Silver, an emergency room doctor in

Phoenix with a passion for nature photography. The persistent physi-
cian, who would join Suckling and Galvin in cofounding the Center
for Biological Diversity, had demanded protection for the spotted owl
under the Endangered Species Act of 1973, relying on what was then
an obscure provision that empowered citizens to petition the government to enforce strict safeguards for imperiled species. Those safeguards, Silver knew from reading the text of the act, made science
paramount. Economic concerns, jobs, progress, and development—
including logging—had to take a backseat under the law when extinction was at stake.

At the time, President Reagan had made opening public lands
to timber companies and other extraction industries (oil, gas, gold
mining, coal, and more) a top concern. Dr. Silver's petition compelled
the Forest Service to shift that focus and determine what threatened
the owl and what protections it might need. And to do that, the Forest
Service had to figure out how many owls were left in the Gila. The
downy birds with the huge dark eyes are elusive nighttime predators,
notoriously difficult to track and count, so the Forest Service adopted
a technique in which crews would walk through the tall grasses and
beneath stands of ponderosa pines and firs, imitating the mournful,
four-note call of the spotted owl, then listening for a reply in the
darkness. Responses could be rare, but sometimes the owls would
alight on a nearby branch and stare down at the strange, featherless
creatures that had called to them; some of the best callers could hold
out dead mice and get the owls to swoop out of the trees and seize
them with strong, sharp claws. These survey crews called themselves
hooters. Surveying was a low-paid summer job, undertaken only reluctantly by the Forest Service and viewed with suspicion by the loggers and ranchers who worried that the lucrative public timber supply
might be closed to them, but the work had its rewards: brilliantly
clear starlit skies; nearly mystical encounters with the ghostly rap-
tors; a sense that something important might be happening, a shift

in the way the forest would be treated, if only the hooters could find enough owls.

Peter Galvin, working toward an undergraduate degree in conservation biology at Arizona's Prescott College, hired on that summer as a contract employee directing a survey crew operating out of Catron County, New Mexico, near the heart of the Gila. One of his hooters was Kieran Suckling, a wild-haired neo-hippie philosophy grad student and former anti-logging activist with Earth First! Galvin had also done a stint at Earth First! and had chained himself to a tree or two in the Pacific Northwest during the timber wars there (a biographical note he somehow hadn't mentioned to his Forest Service colleagues). He recognized a kindred spirit in Suckling, and they soon became roommates, beginning what they call a "conversation" about nature, man, and extinction that has continued for twenty years and has become the guiding force of their Center for Biological Diversity.

Of the two men's journeys to the Gila, Suckling's seems the more circuitous, although his affinity for animals has been apparent since he was ten. He chose Francis as his confirmation name in the Catholic church, after Saint Francis of Assisi, the twelfth-century monk and patron saint of animals and the environment who was said to have preached to the birds and to have negotiated a truce between a wolf and some angry townspeople. Francis's position that all life was sacred and worthy of protection sounds more than a bit like deep ecology; today, as Suckling sees it, Francis would probably be called a radical. Or perhaps an outlaw.

The son of a civil engineer, young Kieran moved every few years as his dad went about building power plants, pulp mills, and other major infrastructure projects—the sort of projects that tend to make environmentalists cringe. They lived in Nevada and Peru, and in the shadow of an oil-fired power plant on Cape Cod. A nuclear power station in South Korea would have been next if his parents had not split up. He and his mother stayed in Cape Cod, where he attended high school in Sandwich and toyed with the idea—not for the last

time—of rejecting everything his engineer father stood for by attend-
ing culinary school, for he loves to cook, and he's good at it. Instead,
Suckling eventually decided to seek a four-year degree, initially at
Salve Regina College in Rhode Island, which had only recently gone
coeducational and was overwhelmingly female. The strict rules of the
Sisters of Mercy who founded the school in 1934 still held sway despite (or perhaps because of) the advent of men, and Suckling found
himself invited to leave and not return after being apprehended in
the women's dormitory one morning. He landed next at Worcester
Polytechnic Institute as an engineering student, back on the path his
father and older brother had taken before him. Worcester was an alternative school, unusually for an engineering college, and Suckling
admired its focus on project-based learning and instruction in the
humanities intended to produce well-rounded engineers. The humanities courses, particularly the philosophy electives, eventually entranced Suckling far more than the study of lines, angles, and stress
points. He had thrown himself into the classics of philosophy several
years earlier when a cousin he admired, a former missionary in charge
of a seminary on the outskirts of Boston, told him the seminary was
closing and he could take home anything he wanted. Suckling carted
off the library's entire philosophy collection and began reading it one
book at time, even though at age fifteen he understood only a small
fraction of the ideas. The seminary's well-worn cloth-bound volumes
of Heidegger and Kant and Plato still occupy significant space on the
crammed bookshelves in his home office in Tucson.

Suckling made it all the way to the last quarter of his senior year
at Worcester Polytech, then abruptly quit without obtaining his engineering degree, although he had very little work left to do in order to
receive it. His family was shocked, as much by his unusual explanation as by his decision: "I just felt that if I had that degree, it would
always be there. You could always fall back on it, get a job. It's kind
of like a scar. It's with you the rest of your life, and you can't take it
back. You'd always have that backup." He practically spits out the

word "backup" when he recalls this cusp, appalled at the very idea of it, this brush with the life and label of an engineer causing a mental shiver at what he might have been if he hadn't followed his heart and fled. Years later, when he would be living in a frigid log cabin, subsisting on old rutabagas and spending what little money he had on faxed press releases about the Forest Service's latest environmental travesty, he would never have to worry about the lure of creature comforts dragging him toward engineering, a steady paycheck, and a life of compromise. But his opponents—governments, corporations, risk managers, and politicians who always have an offer, a compromise, a backup, in their hip pockets—would have to deal with a guy who didn't believe in fallback positions.

With engineering banished for good, Suckling opted for a proudly impractical philosophy degree, undergraduate and then postgraduate (everything but the dissertation—still outstanding after a dozen years). He focused on the continental philosophers, phenomenology, and deconstructionism, and their ideas color his environmentalism to this day, as he thinks and writes on the paradoxical relationship between linguistic diversity, which is vanishing at an alarming rate, and biological diversity, which is also disappearing. He doesn't have to worry about the dreaded career backup with those sorts of intellectual inquiries. "With a philosophy degree, there's nothing to fall back on; it's like casting yourself to the wind," he says. Now pride, sheepish but real, colors his voice. "You get that degree, they kick you out the door, and you have to figure out what you're going to do with your life. You've given up the safety net."

After he left school he took off without much of a plan, spending months driving around the country with two friends, camping in national parks, living on very little money, sleeping in a rattletrap car, grabbing ears of corn from roadside fields and eating them raw. (An amused farmer strode out of one of these fields and informed Suckling and his pals that they had been munching on animal feed corn; the farmer handed them some human corn and sent them on their way.) He settled in Missoula for a time, worked at odd jobs, then moved on, eventually landing at an Earth First! rendezvous in New Mexico, where the radical environmental advocacy group inspired by Edward Abbey's *The Monkey Wrench Gang* had assembled volunteers to protest a national forest timber sale. This was a natural for Suckling: He had been protesting things since he was a kid, beginning in 1979 with his refusal to participate in a mandatory high school rally that was being passed off as a spontaneous student demonstration over the Iranian hostage crisis. He was as upset by the crisis as the next guy, but when his teachers started handing out signs they had made to look like student art, he was outraged. He staged a counterprotest against the "Soviet-style state-sponsored rally" and was promptly suspended. He met a similar fate at the New Mexico Earth First! event, where he was arrested for sitting down in front of a logging truck in an unsuccessful attempt to block its entry into the forest. But the authorities had done him an unintended favor: At the local jail, he met and fell in love with a woman who, as luck would have it, had accepted a job as one of Peter Galvin's hooters, and she invited him to come down to the Gila to see about joining the crew with her.

Galvin's journey also began in Massachusetts, in the town of Framingham, west of Boston, where he was raised in a household suffused not with engineering, but with discussions of social justice, man's inhumanity to man, and classic liberal politics. His grandfather was a Yiddish scholar and his father a devoted Thoreau aficionado, and Galvin's youth was marked by family debates about man's and God's law, about how it was that the Nazis could rise to power, and why the rest of the world could allow the Holocaust to occur. Why didn't Franklin Roosevelt bomb the trains, young Peter wanted to know, and stop the shuttling of Jews to the concentration camps? Why didn't the European powers act sooner when they knew the atrocities Hitler had set in motion? Why didn't more people speak up in the face of evil? These are searing questions for anyone to consider deeply, and the fact that he took them seriously as a teenager presaged the life of adult activism to come, the outrage that simmers beneath a soft voice and outward gentleness. Still, these were academic questions to a boy born in the 1960s to middle-class comfort and privilege. Those horrors seemed to him long past, of a different world, for certainly there could never be another Holocaust.

Such life-and-death questions pondered from afar suddenly cut a lot closer to the bone in 1979, when Peter was diagnosed with testicular cancer. There was nothing academic about listening to his divorced parents argue over which doctor they should trust with their son's life—the one who advocated a series of major, painful surgeries, or the other who would rely primarily on a difficult course of chemotherapy. Each treatment had its own risks, upside, and downside. There were no guarantees; just terrifying discussions of mortality rates and survival rates. And it was Peter who had to make the call in the end. He was fifteen years old.

Galvin opted for three major surgeries, and after a prolonged illness, he made a complete recovery; he has been cancer-free ever since. But his brush with major illness and mortality inevitably altered him. The former captain of the junior varsity basketball team, the boy with the sunny disposition, became withdrawn and solitary, spending long hours holed up in his room reading; his team and his old pursuits, the activities that once had been the center of his days and life, were no longer of interest. He had made a promise to himself, to God, to something—that if he survived the cancer, if he got his future back, he would live a purposeful life, he would find a mission, he would try to make a difference. A hopeful cynicism is how he describes the new outlook. He found his mission while studying biology in 1983 at Lewis and Clark College, and he learned what wildlife biologists and other scientists had gradually begun to realize: Planet earth was experiencing a major "extinction event." Life was dying everywhere, in unprecedented numbers, at unprecedented speeds. To him, as had happened so long ago with the death camps, it seemed no one was doing anything about it.

Extinction itself is neither extraordinary nor unexpected. Mostly invisible (to us), extinction is a regular, natural event—the environment changes and the recipe for life changes with it, which is why some species adapt and survive, and the rest die out. Only the *rate* of extinctions varies,[2] making earth's story, at least for the last billion or so years for which evidence still exists, a cyclic balance of extinction and renewal. That story is written in the fossil record, buried deep in the earth and in the beds of ancient and vanished seas, and it reveals that the normal cycle of gradual extinction has been broken five times in the last 500 million years, when massive catastrophes caused major global extinction events. The most devastating, the Permian-Triassic extinction event, occurred 250 million years ago, when the Earth's landmass was largely contained in one supercontinent, now called Pangaea. Ninety-six percent of all marine species and 70 percent of all land-dwelling species—plants as well as insects and vertebrates—died out, leaving behind only fossilized shells, bones, leaves, and imprints in sandstone. Even those species that avoided extinction were nearly wiped out, suffering a 90 percent mortality rate, so that, overall, the total number of individual organisms on earth was reduced by 95 percent. This is when the ancient trilobites, the sea arthropods that once were the dominant form of animal life on earth, vanished, along with countless other creatures of every shape and size. There are several theories about what triggered this greatest of mass extinctions—catastrophic volcanic eruptions, a series of enormous meteor strikes, or releases of toxic hydrogen sulfide gas from deep in the oceans. Any of these could have brought about ruinous changes in the atmosphere, the climate, and ocean currents and chemistry. Because this happened so long ago, the evidence is insufficient to pinpoint the cause (or, more probably, the combination of causes), but that it occurred is certain. The story of life laid waste, of primordial forests dying en masse, can be read like a book, its fossil layers so deep and thick with ancient corpses that the Permian-Triassic event is often referred to as "The Great Dying." If whatever caused "The Great Dying" occurred in 2009, and humanity suffered similar mortality while avoiding total extinction, it would be as if the current world population of 6.7 billion had been reduced to 335 million—basically where the human race stood 1,000 years ago at the end of the Dark Ages.

The other extinction events were not as severe, but neither were they walks in the Jurassic park. The most recent completed extinction event—the Cretaceous-Tertiary, 65 million years ago, believed to have been caused by an asteroid collision—drove half of all species then living into extinction, including the dinosaurs. It had the additional effect of opening the way for the rise of what was then a minor and barely surviving category of creatures, the mammals. The sixth major extinction event, the Holocene, is not complete: It is taking place now, moving far more swiftly than any of the extinction events before it. Some biologists are predicting the disappearance of half the planet's species in the next 100 years. (Previous extinction events were spread out over hundreds of thousands of years.) This is the first extinction event for which there is no natural cause. It is, Peter Galvin was horrified to learn, strictly man-made.

Galvin came to believe that the world was experiencing nothing less than a new global holocaust, a term which is not used lightly in his family, but which he believes to be appropriate—a holocaust against nature, entirely caused by humanity's outsize imprint on the planet and the environment. It poses a cruel irony, he says, and a great moral quandary, because mankind is at once the executioner of other species, and standing right along with them on the gallows, awaiting the noose just like everyone else.

Galvin dropped out of Lewis and Clark in favor of several years of direct action and protest with Earth First! and other environmental groups, feeling compelled to act and to speak out, but uncertain that chaining himself to trees with unbreakable Kryptonite bike locks was really accomplishing much. He had been in elementary school when a previous generation of protesters helped end the Vietnam War, but the old activist tactics weren't stopping the clear-cuts, the strip mining, or the utter refusal to protect endangered species that began in the 1980s, when Ronald Reagan claimed a mandate to gut regulations and give free enterprise free rein. If there really was a new global holocaust in the making, it seemed Peter Galvin was going to need more than a bicycle lock and good intentions to help avert it.

So he returned to school to complete his degree and his training as a wildlife biologist, transferring his credits to Prescott College because of its reputation for hands-on, rugged, outdoors natural sciences and its focus on student-designed independent studies. Galvin had loved owls since he was a kid, regularly clamoring to visit the Boston Museum of Science so he could see Spooky, a much-loved great horned owl who lived at the museum for a record-breaking thirty-eight years. Owls are high-level predators and fairly ubiquitous in nature, and therefore they make a great index species, as their health in a population often mirrors the overall health of an ecosystem. If owls are thriving, the theory goes, the food chain is intact, top to bottom, and the whole ecosystem is likely to be thriving, too.

When it was time for field study, Galvin chose the Greater Gila Region, which straddles the New Mexico–Arizona border. He was attracted by the land's beauty and remoteness and the opportunity to study an owl species that was clearly in trouble, as its habitat was systematically being cut down. He moved into a tepee near Luna, New Mexico, and contacted the local office of the U.S. Fish and Wildlife Service to see what the staff could tell him about the spotted owl— scientific research, population studies, habitat maps, whatever they had. One of the office workers groaned and said, "Hey, we just had a Freedom of Information Act request on the spotted owl from Dr. Robin Silver over in Phoenix. You know him? No? Well we just sent him a shitload of documents. Why don't you call him?" The worker, eager to avoid another round of time-consuming paperwork, provided the number.

Silver was wary at first, but then became intensely excited when Galvin explained he was studying the owl for his undergraduate degree and had just moved to the Gila region. Silver asked, "Can you get over here tomorrow at three?"

Galvin made the five-hour drive to Phoenix in his old VW microbus and found a dark, barred house awaiting him, and—once the doctor cautiously admitted the long-haired young man on the doorstep—a Spartan interior, except for the spectacular nature photography. Silver explained the barred windows: his opposition to logging had brought him several death threats. It didn't take long for Galvin to determine that Silver was completely dedicated to nature photography and environmental causes. The house was barely furnished. The refrigerator had many rolls of film stored inside but very little to eat. He later learned that Silver, though he made a doctor's generous salary, lived on only $25,000 a year, channeling the rest into activism. Galvin, Silver decided, would be his next project—he agreed that day to become Galvin's patron.

Silver would pay all Galvin's costs for studying the owl—driving, phones, food, living expenses—and in return, Galvin would help with endangered-species petitions on the spotted owl and other creatures who had, at that time, only Silver for a part-time defender. Galvin, who was chronically broke, drove back to his tepee hardly able to believe his good fortune. When he got back home, a message was waiting. The Forest Service had heard he was studying the spotted owl. Would he like to run a crew of hooters? Galvin was stunned by this run of good luck: He would be paid by the government to study the owls and, in effect, to gather the information he intended to use *against* the government. Galvin remembers thinking, "Huh, someone's paying me to sit here in the forest, which I love to do, and imitate owls. This is the greatest job ever!" He got paid thirty-five dollars a day—a long-haired, long-bearded guy in a tepee, contracting with a very straitlaced federal agency he believed to be in the pocket of the timber industry, and he had never been happier.

A few weeks later, an old friend came by to accept his invitation to join the owl crew, and she introduced her new boyfriend, Kieran Suckling. Galvin had another hooter—and the partner he never knew he needed.

they never saw it coming

Peter, Kieran, and Kieran's girlfriend ended up all moving into a rustic cabin—no electricity, no heat, seventeen miles down a dirt road to nowhere—the sort of place that, in winter, you never knew if you were going to make it to your destination, or run into a snow squall that left you stranded. The car was always packed with food, water, and camping gear, just in case.

The cabin sat just outside Luna, New Mexico, in an area known as Centerfire Bog, with a view dominated by the distant 11,000-foot Escudilla Mountain. This is where Aldo Leopold learned to think like a mountain, and where the first officially protected wilderness area in the United States, named for Leopold, is located. It was stark and gorgeous and paradise to Galvin and Suckling, though they almost lost the cabin in a heated argument with the landlord on the issue of rent, which was seventeen dollars a month. Suckling said he had been told thirteen dollars, but for once he decided not to stand on principle (although in those days, having four dollars in his pocket was never a sure thing). They stayed there for the next five years, taking their first, tentative steps in creating a new sort of environmental organization. When Galvin explained his and Silver's work on the endangered-species petition for the spotted owl, Suckling's immediate response was, "Why just the owl? Let's look at everything."

This simple and now seemingly obvious thought—that five or a dozen or 100 endangered species identified in a vast natural region could have far more impact than just one owl—has been their guiding principle ever since. So the research began, in the field, at forest service offices and fish and wildlife stations, and at the University of New Mexico library. Kieran would hitchhike there, sleep in the bushes until the library opened, then spend the day poring over microfiche and congressional reports and dispatches from the forest service, as he and Peter built case after case, slowly, steadily, and overwhelmingly—biologically, legally, and politically. Suckling describes himself as an inveterate gatherer of documents and details—he has been building an immense database of species since those early days, before he had a computer, though now it resides on the laptop he carts everywhere, and he tweaks it many times a day. "Harnessing the power of OCD," he says. Casual acquaintances often mistake this remark for a joke, but it's not: He says he really is obsessive-compulsive.

The Greater Gila Region was and still is a sampler of threatened plants and animals, a situation exacerbated by the fact that the feds flouted the Endangered Species Act almost from the start, and no one had taken the trouble to complain. Suckling and Galvin began to assemble the information they needed for a passel of Endangered Species Act petitions, starting with one for the Southwest willow fly-catcher, a small, rare bird with a distinctive trill—"fitz-bew"—that lives throughout southwestern desert streamside habitats. Saving the fragile riparian lushness that springs up around flowing water in the desert has been one of their longest battles.

Galvin next wangled a fish and wildlife contract to observe the
desert-nesting bald eagle—a subspecies of America's national bird
whose population had dwindled. He would spend ten days in a bird-
watchers' blind behind a juniper tree, then take four days off, then
return for ten more days in the blind, charting the eagles' feeding and
hunting habits and their technique of stealing food from other hunting birds. He witnessed their aerial courtship display, talons locking
as the birds plummeted through the air. Galvin still ranks it, fifteen
years later, as one of the most amazing sights he has ever beheld.
Owing to overarching protections of bald eagles everywhere in the
country, the desert-nesting eagle was safe for the time being, but Galvin's research would be crucial in years to come, when the administration of President George W. Bush tried to remove all protections for
the bird.

Other species they studied and would eventually seek to protect in
this first flurry of activity included the cactus ferruginous pygmy owl,
the northern goshawk (a ferocious hawk prized for falconry in medi-
eval Europe), a parsley-type plant called the Huachuca water umbell,
the Canelo Hills ladies' tresses orchid, the Huachuca salamander, and
the loach minnow and spikedance—two fish which had already been
listed as endangered, but which were receiving no habitat protection.
There were 150 species in all on their list, each threatened by timber
sales, developments, or road projects of one sort or another. It took
more than two years to assemble the information, and they figured
that those plants and animals would be enough to fuel a decade of
litigation, with petitions filed serially to keep the government off balance. The cases would reach far beyond the Gila's boundaries, because many of the species also existed elsewhere in the country. Even
then, in their cabin off the grid, they were laying the groundwork for
a national, not a regional, environmental organization.

They would earn meager salaries during the summers on the owl
crews and then, when the hooters were idle, survive on unemployment insurance and occasional grants from Robin Silver, allowing

them to continue their research. Suckling used some of his old engineering talent to jury-rig a solar panel to power a fax machine—this was before the ubiquity of e-mail—and that became their primary link to the outside world. From the fax, their first environmental alerts and announcements went out to the press and to a tiny donor membership they had assembled by 1992 for their two organizations at the time: Galvin's Friends of the Owl and Suckling's Greater Gila Biodiversity Project.

The forest service's original goal for the hooters was to find as many owls as possible, because this would mean that little or no conservation effort was needed. The forest service could continue business as usual, allowing private timber companies to purchase cut-rate logging rights in publicly owned forests, and all would be well. But when the hooters could find barely any owls in the vast forest, the goal changed: Finding no owls would mean logging could proceed. Timber sales in national forests would be blocked only in areas where owls were found. The birds needed many acres of dense conifers and undergrowth in order to protect their nests, seek out their prey, and hide from other predators. So whenever the survey crews found a mating pair of owls, a 2,000-acre circle was supposed to be drawn around its nest, and that area would, for the most part, be off-limits to logging or other disruptive activities. No owls in a tract the timber companies wanted meant nothing was in danger, and therefore the chain saws could fire up.

Somehow, Suckling and Galvin found, the deck kept getting stacked against the owls: The hooters' reports of owl sightings were downplayed, explained away, or ignored. The logging would continue, even when they found an owl in an area slated for a timber sale. With only a few thousand Mexican spotted owls believed to be still alive, such cavalier decisions infuriated Suckling and Galvin. But when they complained to their supervisors, they were told to mind their own business. Their job was to hoot. Their superiors would handle policy.

Then, during a visit to a forest service office in Albuquerque, a sympathetic staffer mentioned that he had been looking over a map that showed known spotted owl nests in each of the eleven national forests in the Southwest—more than 900 nests. The map was carefully kept from the public; Suckling and Galvin knew it would be a gold mine of information for them. The staffer showed it to them for a moment, then left it sitting unfolded on his desk. An unspoken understanding passed between him and the two hooters. "Well, it's time for my lunch break. I'll be gone about an hour." And he walked out, leaving them alone in the room with the map. They raced to a copy shop, then returned the original; the map, which would be useful for years, told them the forest service had routinely lied when it claimed that areas being logged contained no owls.

The research-underwritten-by-government phase of their work was about to end, along with their hooting, as an inevitable conflict arose between their roles as agents of the government and their position as activists who disagreed with pretty much everything the government was doing in the forest. The impetus came one hot, moonless summer night when a member of the crew got lost in the dark, and wandered into the wrong canyon. All the others had been coming up empty in their appointed areas, their hoots bringing no replies, no sounds of silky wings in the night air, no eyes gleaming in the darkness high in the trees. But the hooter in the wrong canyon found an owl staring back at her with its big round eyes. The next morning, they all pored over maps, trying to figure out where the owl had been, until Suckling jabbed his finger at the spot: "Oh my God, you were in Water Canyon; this is the Water Canyon timber sale." The others stared at him. Water Canyon was going to be one of the biggest sales of public forest timber in the southwestern United States. "That owl is going to be worth millions of dollars in lost timber profits," Suckling said. Clearly, this prospect did not upset him in the least.

It was a Sunday morning. Galvin called the forest supervisor on duty and said, "You've got to stop the sale," then explained why. And the supervisor, to his credit, followed the law and halted logging. But on Monday morning, his bosses reversed the decision and said that the logging in Water Canyon would go through as planned. The allotment had already been sold. When Suckling and Galvin complained, the senior supervisor said, "Look, you should have found that bird sooner. We sold it; they get to cut it. It's too late."

"Well, I understand how the timber company might think that," Suckling responded, "but the thing is, that's illegal. You can't just make up a rule like that. If there's an owl, you can't cut."

But the supervisor wouldn't budge. Despite the unambiguous language of the regulations, he insisted, "You're too late." It was clear the priority was to treat national forests like tree farms for the timber industry. To be fair, this attitude had contributed to the initial creation of the national forest system in 1891, and explained why it was controlled by the Department of Agriculture (whereas the Department of the Interior controls national parks). But conservation was also a critical goal of the forest system, and in the subsequent creation of the U.S. Forest Service by the nation's greatest conservationist president, Teddy Roosevelt. In the Gila, Galvin and Suckling decided, the forest service had lost sight of this part of its mission. So they had a choice. They could shut up, bide their time, and keep making a meager living conducting owl surveys while they continued their species research on the side. Or they could go full time into the nonpaying business of saving species from extinction. Suckling told his partner, "It's time to put out our flag," and Galvin agreed. They drew up a map of the owl location in Water Canyon, brought it to a local newspaper, allowed themselves to be quoted on the record notwithstanding a nondisclosure agreement they had signed to get their jobs with the owl survey, and amid considerable press attention, the forest service had no choice but to cancel the big timber sale. It also canceled Suckling's and Galvin's jobs as hooters.

The time had come to put their stacks of files and long months of research to work. First one, then a couple, then a flood of official filings flew out of Centerfire Bog: petitions for listing threatened and endangered species, formal notices of intent to sue, demands for setting aside critical habitats, and objections to a host of timber sales, zoning variances, and road projects, aided by the secret government map of owl nests. Every time the forest service published the required public notice of a proposed timber sale, there would be Suckling and Galvin with their sheaf of papers explaining why it would be illegal to touch a single one of those trees because of an owl, a flycatcher, or some plant no one had heard of before. It was as if a dam had broken; two years of research, frustration, and trying to work within the system were giving way.

Soon they became notorious in their adopted home, Catron County, which happened to be a ruggedly beautiful environment, but also one of the most hostile locations for environmentalists in the country. Catron County had become the unofficial capital of the Wise Use Movement,[1] a green counter-initiative that sought to temper Suckling's and Galvin's brand of "keep it wild" environmentalism with policies emphasizing the economic benefits of private enterprise on public lands. Catron County politicians had tried to repeal federal environmental regulations, to seize federal lands, and to force environmentalists to register like sex offenders. Suckling and Galvin received threatening letters; their tires were slashed; a car window was smashed. Someone defecated on Suckling's car in a restaurant parking lot. Although a few residents discreetly slipped them contributions now and then, others made it clear that they didn't appreciate two outsiders coming to the Southwest with their East Coast attitudes, trying to put the forests in a sort of lockbox. "They're trying to shut down the West," one angry rancher said at the time. "That's the agenda—locking it all up." Suckling and Galvin responded by changing their letterhead to feature a padlock as a logo, accompanied by the phrase "Two Guys from Massachusetts."

The *Albuquerque Journal* ran a front-page profile of the two in September 1993—the headline, "Going for the Throat," was based on a quotation from Galvin about "going for the jugular" in environmental cases. The newspaper put their bearded, shaggy visages on page one, too. They later found that the front-page photo had been clipped out and posted on a bulletin board at the Catron County courthouse. Someone had drawn a bull's-eye over their image. The forest service officials no longer treated them as an unimportant nuisance but perceived them as a genuine threat, and they were not above using a few guerrilla tactics in order to further that impression. One Thanksgiving Day they spent several hours faxing dozens of files—a notice of intention to sue and supporting documents—to a forest service office, just so officials would come in the next day and see that their opponents were going after them even on a holiday. "Nothing like getting up in the morning," Galvin told reporters at the time, "and knowing you've raised the forest supervisor's blood pressure fifty points."

Their weapon of choice in those early days was (as it is today) the Endangered Species Act of 1973. No other organization has used it more often—or pushed its mandate so far, beyond merely protecting individual species to rescuing and preserving whole forests, marine ecosystems, oceans, rural landscapes, and, if the center's strategists have their way, the global climate. Whether such an expansive application of the Endangered Species Act is cause for celebration or condemnation has long been a matter of debate and the subject of years of litigation. But few on either side doubt the act's potential for changing the world, one way or another, which is why it has endured, and why it has made so many enemies. Admirers worldwide call it the most powerful and effective environmental law ever conceived, the crown jewel and also the pit bull of the environmental movement. Warren Burger, the conservative chief justice of the United States, writing the majority opinion in 1978 when the Supreme Court elevated survival of the lowly but endangered snail darter fish above a multimillion-dollar Tennessee Valley Authority dam, called the act "the most comprehensive legislation for the preservation of endangered species ever enacted by any nation."[2] It was created to do one thing single-mindedly—protect wildlife from unnatural extinction, which in essence means protecting wildlife from man—and to make that mission secondary to no other consideration. The works of man—his factories, his dollars, his bulldozers and chain saws—all must give way when extinction is on the line. This priority is justified, Chief Justice Burger explained, because a road or a truckload of logs or a dam has a specific dollar value that can always be replaced or spent elsewhere, whereas the value of an entire species is "incalculable." Once a species is gone, it's gone forever. The act, then, was crafted as the legislative embodiment of the "precautionary principle," designed to err on the side of protection. (By contrast, most government programs—the Food and Drug Administration, for instance—are designed in many cases to err on the side of commercial interests. That is one reason so many harmful drugs over the years, from thalidomide to Vioxx, have been initially presumed safe for public consumption, only to be recalled amid episodes of birth defects and deaths.)

This precautionary approach has worked for endangered species. A majority of the plants and animals placed on the endangered list for more than ten years have recovered or stabilized, and almost all have avoided extinction. The longer a species has been listed, the better it has fared—that is why most recovery plans for endangered species last forty-two years or more. On average, those species listed in the early days of the law (or under its more limited precursors) have been the most successful of all; they include the bald eagle, recovered from a low of 417 mating pairs to 9,789; whooping cranes, recovered from fifty-four to 513 birds; and the gray whale, which doubled its population in thirty years, to more than 26,000. Of the more than 1,800 species listed under the law as endangered or threatened, only nine are known to have gone extinct once officially protected. The National Research Council estimates that at least 165 currently living species would be extinct now without the act's protections.

The Endangered Species Act is by no means universally admired, however. It is reviled in some quarters as an abusive and unreasonable regulatory burden that interferes with private property rights and the sensible administration of public lands—all to protect a few bugs, weeds, or tiny fish no one has ever heard of or would ever miss, as detractors see it. They believe that the cattle, mining, and timber industries ought to have ready access to public lands; that jobs and human rights ought to come first; and that the Endangered Species Act therefore poses an unacceptable and unnecessarily extreme bar to development and commerce. The law's most vocal critic, Richard Pombo of California— rancher, six-term congressman, real estate developer, and advocate of radical property rights—built his political career running against the Endangered Species Act as if it were a living, breathing campaign opponent, repeatedly attempting to gut its protections. Pombo, who was renowned on Capitol Hill for his black Stetson, his ostrich-skin cowboy boots, and his close ties to the disgraced former House majority leader Tom DeLay, also tried to open up the protected arctic wilderness to oil drilling, sought to bankrupt the endangered-species budget by forcing government to pay individual property owners millions to protect imperiled species, and proposed selling fifteen major national parks for private development. His inaptly titled legislation, the Threatened and Endangered Species Recovery Act, would have ended all species habitat protections by 2015, replaced scientists with political appointees in deciding which species to list, and required taxpayers to pay corporations hundreds of millions of dollars just to obey the Endangered Species Act—the equivalent of paying drivers to wear seat belts. The bill passed the House but stalled in the Senate as the Center for Biological Diversity led a huge media and lobbying campaign to expose the effects of the innocuously named legislation. It finally died once voters ousted Pombo—and the Republican Party's congressional majority—in 2006.

It is safe to say that if the Endangered Species Act had been introduced in Congress in the political climate of the early twenty-first century rather than in 1973, it would have no chance of passing. Yet it was utterly noncontroversial and nonpartisan when it passed, viewed as moderate and middle-of-the-road. It was cosponsored in the Senate by a group of Democrats and Republicans; it passed on a unanimous Senate vote and with only four "no" votes in the House; and it was signed by President Richard Nixon, a Republican who lavishly praised the law and his program of companion ecological measures as part of an "environmental awakening," a "commitment to responsible partnership with nature," and a rejection of "cavalier assumptions that we can play God with our surroundings and survive."

The act revolved around a few simple ideas few found objectionable: Americans cherished nature and did not want animals to become extinct; therefore, it should be illegal to kill endangered species or destroy habitats critical to their survival. The near extinction of America's national symbol, the bald eagle, helped galvanize politicians into taking this action, and they were also impelled by a growing environmental movement that brought 20 million Americans to rallies on the first Earth Day, April 22, 1970. An unprecedented wave of environmental laws then swept through Congress: an improved Clean Air Act, the Clean Water Act, the National Environmental Policy Act, the Marine Mammal Protection Act, the creation of the Environmental Protection Agency, and the Endangered Species Act. Nothing like them had ever been passed before, and nothing like them has been attempted since.

The effectiveness of these laws depends on their enforcement by the president and his cabinet and agency heads, and performance in this regard since 1973 has ranged from uneven to unenthusiastic to lax to downright hostile, depending on who has occupied the White House. The author of the Endangered Species Act, Senator Harrison Williams of New Jersey, anticipated that, once the rosy glow of doing something good for the environment had passed and the self-congratulatory press conferences were forgotten, there would be great reluctance within government to rigorously enforce environmental legislation whenever high-minded principles collided with the interests of powerful corporations. Inevitably, it would be argued that jobs, the economy, and the interests of hardworking citizens were being made subservient to a few scrawny owls. Elected officials and their appointees would find it hard to resist the temptation of ignoring the act's provisions. This is why the Endangered Species Act, like several of the other environmental laws adopted during that era, grants general "standing" to any U.S. citizen anywhere in the country to petition and sue the government over any endangered species anywhere. The goal of this unusual provision—most laws do not permit such broad standing—was to sweep away legal and technical obstacles to holding government accountable. Although it has been criticized as "government by litigation," the record of how and why species have been put on the endangered list makes it clear that the list of protected species would be a great deal shorter, and the list of extinctions a good deal longer, without the citizen-suit provision and the power of court orders. When the Senate voted unanimously for the act in 1973, this was what was intended. The forest service hates it. The Center for Biological Diversity lives by it.

Yet, for all the power of the Endangered Species Act to hold government's feet to the fire, the first sixteen years of the act's existence saw relatively little use of the citizen suit. Its most famous early invocation was the case of the Tennessee Valley Authority dam in 1978—a case that the snail darter won and the Tellico Dam lost, until an amendment to an unrelated bill cleared Congress and exempted the project from the Endangered Species Act. More than a decade would pass before the next big battle was waged in the Pacific Northwest—the effort to protect the northern spotted owl, a relative of the bird Suckling and Galvin wanted to save. That court battle also curtailed logging, but it created enormous controversy, a backlash, and ill will, as the Pacific Northwest timber industry was much bigger, more powerful, and employed more people, than the industry in the southwestern forests. The northwestern timber industry had been declining for many years, largely because of overseas competition, but environmentalists became a convenient scapegoat for the industry's woes and for workers with no jobs. Spotted-owl dolls were hanged in effigy at sawmills, and bumper stickers proclaiming such sentiments as "I Love Spotted Owls—Fried" became commonplace. The perceived enemies were those damned environmentalists with their lawsuits.

Mainstream environmental organizations had almost always made their greatest strides—such as the campaign against dams in the Grand Canyon— not through lawsuits but through lobbying and public appeals. That was thought to be the best way to build broad support for environmentalism, working within the system rather than suing it, and the legal battles in the Pacific Northwest and the rising tide of antienvironmentalism they engendered seemed to reinforce this notion. But 1981 brought a change. From the start of his presidency, Ronald Reagan sought to cut environmental regulations as part of his overarching disdain for government; and his department heads simply stopped enforcing those laws he lacked the power to eliminate. Reagan wore his antipathy toward environmentalism like a badge of honor. He told America, "Trees cause more pollution than automobiles," an outrageous lie. He also opined, "You've seen one redwood, you've seen them all"; this is probably why he put a timber executive in charge of national forests.

The leader of Reagan's attack on environmentalism was James Watt, his first secretary of the interior, one of the legal thinkers behind the Wise Use Movement. Watt openly despised the Endangered Species Act and refused to list a single endangered species for 382 days straight—a record that would endure for a quarter century.

Watt's fiery two-year tenure consisted of a string of cuts in environmental funding, enforcement, and regulations, and corresponding policies to open up public forests and lands to oil and gas exploration, grazing, and development. In 1983, when he was criticized over a controversial decision to sell a billion tons of coal to be strip-mined from public lands, he attacked the composition of a federal coal advisory panel by saying, "I have a black, a woman, two Jews, and a cripple. And we have talent." Watt was forced to resign in the wake of that remark. A few years later, he let his unvarnished feelings about environmentalists be known in a speech before the Green River Cattlemen's Association in Pinedale, Wyoming: "If the trouble from environmentalists cannot be solved in the jury box or at the ballot box, perhaps the cartridge box should be used."

Endangered species listings picked up again when the first President Bush took office, and more so under President Bill Clinton (who holds the record for listing endangered species: 527 in two terms). But Reagan's and Watt's influence remained strong in the forest service long after they left office. The affable Reagan, master of the pleasingly false anecdote, who described Watt as an "environmentalist" in the same way he asserted that trees caused forty times more pollution than cars, had succeeded in recentering American politics to suit his right-wing preferences, making what had been considered a politically moderate idea at the time of its enactment—the Endangered Species Act—seem like a creature of the radical far left. Environmental organizations, desperate and dispirited in this political climate, sought to compromise and placate, hoping to hold onto some of their past gains and weather the storm. The in-your-face, take-them-to-court strategy of the Center for Biological Diversity made other organizations nervous. Even Washington, D.C., attorney Brock Evans—a stalwart of the Sierra Club and Audubon Society who had fought to save the parks and forests of the Northwest—wrote to the center and suggested Galvin and Suckling back off from their lawsuits to avoid the ire of the powers that be.

They politely told him to forget it (and Evans later said they were right to do so). As Galvin sees it, 90 percent of America's wilderness has been used up, with no concern for extinctions and without compromise. "Now that last ten percent of American wilderness has some protection, and they suddenly want us to compromise so they can use it up, too? I don't think so." Suckling offers a pointed metaphor:

"This is how we should all be looking at the extinction crisis—Your house is burning down. Do you want a fireman who stops on the way in to cut a deal to save *part* of your house?"

Suckling and Galvin had decided to turn Watt's advice on its head: For environmentalists, the cartridge box was not an option, and the ballot box had stopped working, so the jury box—or, more precisely, the judge's bench—was the only way left to go. The Endangered Species Act had been waiting for years, and though petitions were filed here and there by other groups, the Center for Biological Diversity was the first to systematically bombard the government with petition after petition, suit after suit, until finally its priorities became the government's grudging priorities as well.

The two former hooters launched a flurry of cases in the early 1990s, but the first one to work its way through the long pretrial process and catch fire involved the bird that had started it all, the Mexican spotted owl.

After a four-year delay, Fish and Wildlife had finally ruled in favor of Robin Silver's old petition and put the owl on the endangered list in 1993, but with no protection for its habitat—the forests where the spotted owl nested were still being heavily logged. To Galvin and Suckling, this was an intolerable oversight, as if a government engineer saw a bridge about to collapse but took no action to shore it up. It taught them that there was never a final victory in these cases: You had to get a species listed; then you had to sue to have its habitats preserved; then you had to sue again when the protection was insufficient or the government tried to reduce it. Galvin and Suckling had merged their two groups, Friends of the Owl and the Gila Biodiversity Project, into the regional Southwest Center for Biological Diversity, with the third cofounder and financial backer, Robin Silver, as chairman of its board of directors. But the center had no lawyers on the staff (and, at that time, no staffers other than themselves), so they had to recruit environmental attorneys who would donate their time in the hope that, if they won, the government would be ordered to pay their fees. A single species could require years of litigation, so it was just luck that the spotted owl ended up breaking first.

The case began when Silver, Galvin, and the volunteer attorney, Mark Hughes of the Denver firm EarthLaw, met with officials of the U.S. Fish and Wildlife Service to see what would happen next in the quest to save the spotted owl from extinction. The three environmentalists wanted to know why the critical habitat areas the birds needed had not been set aside. It would take more than just drawing circles around nests—substantially larger, *contiguous* areas had to be set aside for the owls to inhabit, Galvin, Silver, and Hughes argued. Each time the owls nested and had young, the juvenile birds would be ousted from the parents' territory and they would have to find their own 2,000 acres for hunting, mating, and nesting. Like many birds, spotted owls are intensely territorial, even with their own offspring. The people at the fish and wildlife service understood this, but said they were helpless to do anything about it. That was up to a separate agency, the U.S. Forest Service, which had no interest in setting aside vast areas of forest for the spotted owl.

After the meeting, the activists and their attorney realized they had hit pay dirt. When it came to endangered species, the forest service was required to defer to the fish and wildlife service, not the other way around. The problem was that the two agencies weren't consulting with each other, as required by the Endangered Species Act. The fish and wildlife service was supposed to make a scientific finding—a written "biological opinion"—that would assess how logging and building in the national forests would affect the spotted owl. With this opinion, a comprehensive protection plan could be crafted. Without it, trees were being cut down in great numbers in the national forests, and no one had any idea how this was affecting the spotted owl. In short, they had a slam-dunk legal winner, because each day that went by, the feds were breaking the law.

In December 1993, they filed the required notice of intention to sue the U.S. Forest Service, beginning a ritual that was new to them at the time but would become a regular pattern of events in the future. The notice triggered a mandatory sixty-day negotiation period, but the forest service took no action to settle matters. As promised, they filed suit, and the following June, a prominent jurist in Phoenix, Arizona, U.S. District Judge Carl Muecke, ordered the fish and wildlife service to follow the law and designate a critical habitat for the spotted owl. Had the agency done as the judge instructed, the case would have ended then and there, with habitat preservation for the spotted owl that would probably have been far more limited than Suckling and Galvin desired. But federal wildlife officials in Washington, D.C., did not do as they had been instructed, apparently unimpressed by the two bearded activists living in an unlit cabin and the yokel judge out in the desert. Suckling and Galvin soon received an anonymous packet of memos from their old sources inside the government, and those memos made it clear that the agency had no intention of setting aside critical habitat in any of the southwestern forests.

In short order, those memos were presented to the judge. Suckling and Galvin, like most attorneys in Phoenix, knew something that their Washington-based opponents apparently didn't appreciate: You don't mess with Judge Muecke. He was a no-nonsense ex-Marine who had served in the OSS during World War II. Muecke hunted down Nazi war criminals. He personally arrested Hitler's filmmaker, Leni Riefenstahl. He participated in the liberation of the concentration camps at Dachau and Buchenwald, where he saw firsthand the starvation and abuse of the living, and the mass graves and the ovens that had consumed the dead. The experience transformed him into an impassioned civil libertarian, a legendary liberal in a conservative state, a judge who did not defer to governors, congressmen, or presidents, much less bureaucrats—all of whom he had angered at one time or another. Those memos, as Suckling recalls it, left him "royally pissed."

In August 1995, Muecke issued an injunction that banned all log-ging in the eleven national forests of the Southwest until the forest service and the fish and wildlife service came up with a legitimate biological opinion on logging and its effect on the spotted owl. The governor of Arizona at the time, Fife Symington III (who was later forced to resign after being convicted of bank fraud), railed against the judge, with whom he was already feuding over prisoner-rights lawsuits. Hundreds of loggers drove their semis to Phoenix and sur-rounded the courthouse like a raiding party, blowing their air horns and blocking traffic. But the forest service had caused the delay, not the judge, and nine more months passed before a document labeled "Biological Opinion" was finally filed. It contained a great deal of information, none of which had anything to do with biology or habi-tat; it focused instead on an economic analysis of how the timber industry might suffer because of owl protections. Muecke rejected the report.

A second biological opinion appeared and was rejected for simi-lar reasons. Then in July, after a year under the logging injunction, the forest service issued a press release announcing that logging was going to resume immediately in the southwest forests. The regional forester, Chip Cartwright, then one of the most powerful forestry officials in the country—who had also been in charge of ecosystem management for all national forests—had filed yet another biologi-cal opinion. He faxed it to the judge late on a Friday, after court had adjourned, and said that he had fulfilled the conditions the judge had set for lifting the injunction. The new biological opinion, to no one's surprise, found that the spotted owl faced no danger from continued logging operations. Cartwright called the supervisors of all the forests and told them to get the timber companies back to work.

However, the judge had not approved of this action—the forest service had acted unilaterally. Muecke was furious, ordering an im-mediate stop to the logging and summoning Cartwright to the court-house in Flagstaff, near his home, where the judge was recovering from surgery. Cartwright was facing potential contempt of court charges; someone suggested he might want to bring a toothbrush. At a secret hearing, the judge looked him over and asked why a defendant in a court case felt he had the power to declare the case over and to lift an order of the court. "I have never run into that in my whole life," the judge fumed. Cartwright avoided a contempt citation, but the injunction remained in effect for the rest of the year, as Muecke ordered the parties to lock themselves in a room and work toward a settlement.

This was what Suckling and Galvin had been waiting for: with feds on the defensive, a judge who was one affront away from jailing the director of all southwestern national forests, and logging at a complete standstill in their beloved Gila, they would be locked in a room with their opponents—forest service guys who would have crossed the street to avoid bumping into the bearded enviros on the sidewalk. The goal was to wear them down and it finally worked. On day five of the settlement talks, as Suckling and Galvin recall, one of the Department of Justice's senior lawyers sent in from Washington, D.C., to negotiate apparently reached his breaking point. He abruptly stood up, a strange look on his face, and walked almost robotically around the conference table and grabbed the attorney, Mark Hughes, attempting to throttle him, breaking his glasses in the process. Galvin and Robin Silver, who was the "client" in the case, leaped up to restrain the government lawyer. Later, the official story was that the justice department attorney had "bumped" Hughes, to which the environmental attorney agreed, but only if the term "bumped" included wrapping hands around someone else's neck. The case settled the next day and the center had its first big victory. The forest service agreed to a Mexican Spotted Owl Recovery Plan in all eleven regional forests. The owls would have 4.6 million acres of critical habitat. Logging could resume, but in a much reduced form—always secondary to the health of the owl and other species in the woods.

Suckling and Galvin returned home in triumph, only to be evicted from their cabin in Centerfire Bog. They were too hot to handle, their landlord told them sadly. He didn't want to evict them, but he had to live in that community, which was less than thrilled with the outcome of the spotted owl case.

After a stint in Silver City, New Mexico, they moved to Tucson and soon dropped the word "Southwestern" from the name of the center, eager to build a national organization and reputation. The federal government unwittingly helped by keeping the spotted owl in the headlines, continually cheating, attempting to revise the settlement, and outright disobeying court orders to protect owl habitats. By the time the government was done irritating various judges and appeals courts, logging and cattle grazing in the southwestern forests were greatly curtailed, and the protected critical habitat for the spotted owl had been nearly doubled, to 8.6 million acres.

"We effectively ended the timber industry in the Southwest," Suckling boasts. "And it was their arrogance that did them in. They could have settled things at the beginning, and we never would have gotten where we are today. They never saw it coming."

Today the Center for Biological Diversity occupies a large and somewhat cluttered office inside a former market, which the staff must vacate one month a year when the landlord reclaims the premises to participate in Tucson's annual gem show. Kieran Suckling and Peter Galvin consider this a good deal, however, because the place comes rent-free the rest of the year, and includes some impressive, if rather dusty, giant purple crystals that are too heavy to move. It's also a hard place to find if you don't know where to look—as a safeguard against walk-in hotheads, they've never put up a sign or a logo for the center. The parking lot in the rear is unmistakable, however: the prevalence of bumper stickers featuring the word "save," followed by some critter or natural landmark, is a clear giveaway. Inside the converted market are groupings of couches and coffee tables, a small

kitchen, a library, a warren of desks with low dividers in the cavern-
ous main area of the building, a suite of offices and meeting rooms to
one side, and a steady buzz of activity throughout as staffers pursue
their diverse petitions, lawsuits, campaigns, fund-raising calls, and
media contacts, often long into the night.

In terms of its reach and staff, the center has come a long way
from its first incarnation as two guys and a solar fax machine at
Centerfire Bog. Within two years of moving to Tucson, Galvin and
Suckling had assembled a (barely) paid staff of sixteen and an annual
budget of about $400,000, much of which came from grants from
Doug Tompkins and Ted Turner. But they pulled in $100,000 from
the center's 4,000 members (up from a mere 250 members three years
earlier). Almost the entire budget was churned into litigation costs,
campaigns, publicity, and scientific studies; the staff pay topped out
at $350 a week for Suckling, the executive director, with the over-
all average salary at $250 a week. A substantial number of staffers
lived at the center for extended periods; the combination campsite-
commune-crash pad ambience even included Galvin's old tepee in the
backyard, though the center gradually became more of a professional
operation as the wages slowly evolved into low but livable salaries in
exchange for exceptionally long hours.

Each year since then, the center has grown steadily and extended
its reach. By 2008, it had filed more than 500 lawsuits and won about
90 percent of them. It has won first-time protections for a total of 350
species, 70 million acres of critical habitat, and tens of thousands of
miles of river.

Now it maintains a law office with about twenty lawyers, biologists, and a support staff in San Francisco—home to the federal trial
and appeals courts most favored by environmentalists—along with
satellite offices across the country in Washington, D.C.; Los Angeles, San Diego, Sacramento, and Joshua Tree, California; Phoenix
and Prescott, Arizona; Silver City, New Mexico; Missoula, Montana;
Portland and Bend, Oregon; Chicago; Duluth, Minnesota; Carencro,

Louisiana; and Richmond, Vermont. The center's increasingly active and high-profile climate-change and oceans programs are based in the Mojave Desert on the edge of Joshua Tree National Forest, where the two attorneys who run those programs live in a solar-powered house— and have built a reputation as two of the leading global-warming litigators in the nation. The center's revenue has grown as well, reaching a new high of $6 million in 2007, with expenses just over $5 million— a 50 percent increase over the previous year. Benefactors include the center's 40,000 individual members, the musician Bonnie Raitt, and the Swiss medical device magnate and eco-philanthropist Hansjörg Wyss, whose pledge of $10 million over five years (the most generous donation the center has yet received) is largely responsible for its recent budget bump.

The center still receives considerable scorn and criticism. Fifteen years later, the spotted owl case still remains a sore spot in the Southwest—as many as 3,000 jobs were lost and several timber mills were shut down because of the reduction in logging the center achieved. And the center's latest campaign to exclude all-terrain vehicles from a large percentage of public lands, because of their destructive impact on habitats—a campaign funded with a grant from Doug Tompkins— has aroused considerable venom in the western states, where off-road recreation is common, its supporters are vocal, and the industry is a powerful force. When the *Tucson Citizen* newspaper published a profile of the center (the story described an organization that "brandishes the Endangered Species Act like a blunt force instrument"), the paper's Web site received a flurry of comments, almost all of them negative. "These people would have us living in caves and eating berries but only after the animals got their share," one reader commented. A self-described environmentalist offered this comment: "They are radical and rarely part of the solution. . . . They prefer the fight and to have their own little empire to puff their chests up." Another commenter wrote, "These people love rodents, bugs, and birds more than people. Such inhumanity ought to be mocked. . . . That's why I say pass the owl. Anybody got any good BBQ sauce recommendations for endangered red squirrel?" And there was considerable cheering among readers over a successful defamation suit against the center by the Arizonan rancher James Chilton, who had been profiled in the *Wall Street Journal* after winning a $600,000 judgment in the case.

A jury in Tucson found that photos posted on the center's Web site painted a false image of alleged environmental damage on leased public lands at Chilton's cattle ranch, which actually were lovely and verdant. The photographer hired by the center had been in error. The center argued that the photos were public records, but the state appeals court found that argument to be untimely and upheld the verdict in December 2006. The cattleman, who is also an investment banker in Los Angeles and the husband of a controversial official in the Arizona fish and game department, blasted the center as an extremist, radical organization that wanted to drive ranchers off public lands and out of business. "I had been attacked, attacked, and attacked," Chilton later said. "I had to decide if I was a wimp or a cowboy. I stood up, I cowboyed up like a cowboy should."

Environmentalists, despite occasional squabbles and leeriness about the center's no-holds-barred tactics, tend to be more complimentary, describing the center as a positive force, and crediting it with pushing concerns about endangered species to the forefront of the conservation movement. Peter Bahouth, the former director of Greenpeace, former head of the (Ted) Turner Foundation, and now director of the U.S. Climate Action Network, called the center "fearless," adding, "They have done more to protect nature and hold the government accountable than organizations with ten times their resources."

The center's growth accelerated following the inauguration of George W. Bush, by any objective measure the president most hostile to environmentalism in U.S history. In the absence of congressional oversight or the wholesale intervention of larger, mainstream environmental groups, the center became one of the few consistent checks on Bush administration officials who seemed determined to scuttle environmental laws, alter the science behind them, or simply ignore both and dare anyone to challenge them.[3] The center's investigations and courtroom victories repeatedly embarrassed the White House, as when the center exposed and sued the administration for improperly shelving endangered species on a waiting list called "warranted but precluded"—a legal limbo that has trapped 286 endangered species without providing them any protection. At least twenty-four of the wait-listed species have since gone extinct.

In 2005, the center led a national media and lobbying campaign that stopped Richard Pombo, Congress, and the White House from gutting the Endangered Species Act. A lawsuit brought by the center forced the Bush administration to produce a congressionally mandated scientific report on the impact of climate change; due in 2004, the report was finally made public in 2008, with conclusions that contradicted most of the president's statements and policies on global warming. Then, with twelve state attorneys general and a group of environmental organizations, the center won a landmark case in the U.S. Supreme Court, *Massachusetts v. EPA*, invalidating the Bush administration's position that greenhouse gases were not pollutants and could not be regulated. A short time later, in 2007, the center won another case against the Bush administration, when the U.S. Ninth Circuit Court of Appeals threw out the administration's low fuel-economy standards for sport utility vehicles for failure to consider the impact of greenhouse gas emissions.[4]

And then there was the Julie MacDonald scandal.

An investigation by the center revealed how the administration pressured scientists and falsified data in order to deny or overturn protections for endangered species. The scandal revolved around a deputy secretary of the interior, Julie MacDonald, whose job seemed to be to limit or eliminate protections for endangered species, all over the nation and the world. A farmer and civil engineer from California with no training as a biologist, MacDonald—according to the Interior

Department's inspector general—set out to badger and intimidate biologists in the fish and wildlife service who had found a species or habitat that required protection. MacDonald denied any wrongdoing, but the inspector general, the center, and congressional testimony about MacDonald resulted in a portrait of a political appointee who seemed to rewrite and reverse scientific opinions and data to suit the administration's pro-development, antiregulatory policies. She censored peer-reviewed science that favored protections for endangered species and replaced it with unverified claims by industry that protections were unnecessary for various species. She leaked information to industry lobbyists that aided them in lawsuits against her own department. And she reversed protections that had the potential to affect her own property in California.[5]

Suckling and Galvin uncovered evidence of MacDonald's meddling and her rewriting of science[6] while they were researching the endangered Sacramento Delta smelt, a small silvery fish that was being driven to extinction by excessive pumping of river water for agriculture. The little fish would prove MacDonald's downfall. As she sought to remove the smelt from its listing as a threatened species in order to allow even more river water removal, she had replaced real scientific findings of impending extinction with unsupported claims from farming interests that declared the smelt to be in great shape. The attempt to delist the smelt was rebuffed in court, and Suckling passed evidence about MacDonald to the *Washington Post*. The resulting story created an uproar, and was picked up by all the major media. The Interior Department's inspector general later found that MacDonald had violated regulations and quoted a fish and wildlife deputy director as saying MacDonald was an "attack dog" against endangered species. She resigned in May 2007. The head of the U.S. Fish and Wildlife Service conceded MacDonald had acted improperly, reopened eight endangered species listings and eventually moved to restore protections in seven of them.[7] The center sued, asserting those concessions fell short. At least fifty-five other listings and 4 million acres of habitats in twenty-eight states had been denied protection and needed to be revisited because of an apparent politicization of science.[8]

For all its growth and success, the center remains very much a reflection of its founders' sensibilities, though a new generation of leaders is rising. Galvin is still the master strategist and negotiator, picking the center's fights, expanding its mission to include international cases. He lives in northern California these days, and oversees many of the larger cases on the West Coast, but he speaks daily with Suckling, plotting and brawling with his old roommate, who assumes the role of naysayer, finding reasons not to do nine of the ten new projects Galvin has in mind. "That way, we know for sure the tenth one, the one I can't shoot down, is a winner," Suckling says.

Except for the four years between 2004 and 2008, Suckling has been (and is currently) executive director of the center. He stepped away temporarily from administrative duties only after he and his wife, the novelist Lydia Millet, who is the center's media editor, had their first daughter, Nola. Nola's waifish smile evokes a rarely seen tender side of the intense and sometimes distracted Suckling. Only Nola can get him to put away his cell phone, his laptop, the species database he is constantly tweaking—and his obsession over saving species from extinctions. More than anyone else's, Suckling's imprint is visible in the daily life of the center—in the long hours he (and therefore everyone else) works, in his monkish uninterest in creature comforts, in his preference for hiking as a form of relaxation, in the mostly good-natured tolerance his coworkers display for his chronic inability to keep track of time or appointments, and in his immersion in the science of conservation biology and the philosophy behind it. He has continued writing about the link between vanishing biological diversity and the loss of linguistic diversity in humans. As we diminish and withdraw from nature, and as the extinction crisis mounts, he says there is evidence that we also diminish ourselves, our culture, and our language.

He explains the theory through trees and symphonies. A century ago, Suckling observes, most Americans would walk down a tree-lined street and could, without thinking about it, point out the oak, the birch, the chestnut, the beech, the sycamore. Children knew the difference, and there was nothing remarkable about knowing it. It was simply part of the daily language: "I sat under the old oak and read a book today." Now Americans just see and say "tree." The deeper knowledge of nature, the specific language that separates one tree from another, is now disused and beyond most of us; how many American kids today can tell an oak from an elm? This, to Suckling, is not trivial, but a marker of something larger, a divorcing of humans from the natural world. Or consider, he says, Beethoven's Fifth Symphony, those famous first four notes, dramatic and emblematic, perhaps the most recognizable piece of music in the world. But what audiences in Beethoven's time knew, and what few in the twenty-first century understand, is that those notes, dah-dah-dah daaaaah, have a meaning; they are a reference to nature—those notes mimic the song of a once common bird, the white-breasted wood wren. It was obvious in Beethoven's time, when the experience of nature was far deeper and richer than it is for many Americans today: Listeners heard that refrain, and thought "wood wren." "So what does that mean?" Suckling asks, then answers his own question. "It means the human experience is becoming increasingly impoverished as plants and animals become extinct, or as our lives become so removed from nature that the experience of those plants and animals becomes extinct."

It's clear that Nola's birth has brought Suckling some new insights into Americans' disconnect from nature, though his assessment is more ironic than hopeful.

"A child is born," he muses. "So what's happening in that first year? They're learning to be human. So what do we do when this young creature has to learn how to be human? We surround them with animals. That's the first thing we do, right? The toys and the cartoons and the music and the wallpaper and the books and the nursery rhymes. Everything is animals and jungles and forests. Isn't that bizarre? That we, who are driving so many species extinct, who exclude plants and animals from our lives so radically, at the same time surround these young children with animals at the very time they're becoming human. . . . And then we grow up and say, enough, we're done. Screw nature. What weird schizoid creature would behave that way?

"Really, at the end of the day, I find this very frightening. Think of the experiment we're in now. Humans have been around for hundreds of thousands of years and, for the first time, we're saying, here's an idea: Let's try to live without plants and animals. This is uncharted territory. No humans have ever lived on earth the way we're living on earth. And there's no guarantee that this experiment is going to have a less than horrific outcome. Which, of course, is why the center exists: to cut down on at least some of those horrific outcomes."

the polar bear express

There is a dirt road on the scalding edge of California's Joshua Tree National Park, where the Mojave and Colorado deserts merge in magnificent desolation, and a forbidding preview awaits of what global warming has in store for the rest of the country if it goes unchecked. The road leads to a low-slung house set in a parched terrain of scrub and foothills, the sky bleached white by the sun, solar panels on the roof, and a hybrid in the driveway.

It is a stark, lonely, and beautiful place, and it may well be the nation's most important front in the battle against climate change, though not because it is a carbon-neutral homestead worthy of imitation nationwide. It is also the place where a White House that never backs down got its comeuppance, where a strategy was devised that forced George Bush to admit global warming was happening, where plans are taking shape that could move the country from lip service to action on the most pressing environmental crisis humanity has yet faced.

It is a house where the polar bear is king.

In the polar bear—iconic, charismatic, beloved—the Center for Biological Diversity finally found its poster child, its symbol, the species that got Americans to emerge from their apathy and demand a solution to the greenhouse gases that are heating the globe and melting the polar ice packs. The deadly chain reaction endangering the polar bear is also threatening the future of civilization and humanity, but it was the ferocious and beautiful polar bear that finally seemed to get many people aroused about climate change, while putting the center on every network and front page in the country. The center is determined to use the polar bear to transform the Endangered Species Act into a vehicle for battling climate change. It is a strategy that has enraged the center's critics, flummoxed the White House, and left the global warming deniers in Washington sputtering, as they point out that the act was never intended to be used this way, that the center is engaging in a naked attempt to impose government policy by litigation instead of legislation, and that it's just plain crazy to try to set global warming policy by means of a bear. But the craziest aspect of the center's polar bear project is this: It just might work. If the letter of the law is followed—a big if, given the immense stakes, the powerful forces aligned against the center, and the public's longstanding inertia on global warming—the center wins. If Congress and the president simply obey the law, the polar bear will be protected. And the only way to protect the polar bear is to take on global warming.

"Our hair is on fire on this issue," says Kassie Siegel, whose house in Joshua Tree serves as headquarters for the center's Climate, Air, and Energy Program. One of the new generation of leaders at the center, she has been called three times since 2005 to testify before Congress on global warming, and she has been quoted in more than 1,000 news articles and broadcasts since starting the polar bear project. Her impact on national climate law and policy and her growing national clout have made her a force for change, the center's new eco baron. But she says there is no time or cause to celebrate her rise from obscurity. "The polar bear is running out of time. We are running out of time. This is our shot."

Kassie Siegel lives and works in her desert outpost with the attorney Brendan Cummings, head of the center's oceans program, her partner in law and life. They have offices at the two ends of their home and live in the middle, where picture windows look out on a veritable crossroads for wildlife—bobcats, quail, roadrunners, and owls. Midday exercise requires no stair-steppers or treadmills—just a hike through the desert that abuts their back door. Then they go back to work, back to the legal briefs, the constant correspondence, the phone calls to the tight-knit community of climate litigators that they help lead, the endless federal Freedom of Information Act requests, their main weapon for uncovering the government's violation of environmental laws. They have boxes of documents painstakingly assembled—some of the requests take years to fight—a treasure trove of climate hope and perfidy dating back to the fateful, pivotal, disastrous moment when Ronald Reagan ripped Jimmy Carter's solar panels off the White House roof and killed plans for the United States' energy independence—the moment, Siegel suggests, when the path to global warming moved from possible to inevitable, the worst of the Reagan Revolution's unintended consequences. Those files have produced one revelation after another, such as the extensive war plans the Pentagon drew up for a warmer world with melted polar icecaps, flooded coastlines, and extensive areas of drought— a world the military strategists believed was coming, even as their commander in chief at the time, George Bush, publicly denied any such future awaited. One Pentagon report they found predicts an approaching age of regional wars over vanishing energy, water, and food if global warming remains unchecked; the report is titled "Imagining the Unthinkable." When it came to the military dealing with global warming in the Bush years, there was no denial, no suppression of science, and no expense spared—the only branch of the government for which that was true.

This is deeply depressing to Siegel. Her nightmarish vision: picking up a history book twenty or thirty years from now and reading that the United States, revered as the world's savior during World War II, is to blame for destroying the world as we know it. Future historians will be baffled and our descendants outraged, she fears, because it will be clear that American scientists understood the problem better than any others in the world and had the technology to deal with it through renewable energy and sensible growth, yet the nation lacked the will to see it done. With 5 percent of the world population sucking up a quarter of the world's energy and emitting a quarter of the world's greenhouse gases (which amounts to twenty metric tons a year of carbon dioxide for every living American man, woman, and child), America bears the brunt of responsibility for global warming. Politicians do not like to say this, but it is irrefutably true. "We are rapidly reaching the point of no return on climate change," Siegel laments. "Rome is burning, and they are fiddling. . . . So, yes, this consumes me. It's what I eat, sleep, and breathe. Because we've got to change course, and we've got to do it now."

So in that house in Joshua Tree, life revolves around work. Five more years, tops, and it will be too late to avoid disaster, Siegel believes. Even in her spare time, she remains obsessed with altering the global warming trajectory, serving as one of a thousand volunteers in the Climate Project, presenting Al Gore's slide show from *An Inconvenient Truth* to community groups in some of Southern California's most conservative, least receptive enclaves. Her disarming, easygoing style and her encyclopedic knowledge of climate science and law win over a surprising number of converts. One real estate attorney in San Diego was so impressed by her presentation that he arranged for her to speak to a group of developers, normally the last people on earth who want to be lectured about global warming.

Like many of the staffers at the center, she and Cummings describe their work as a sixteen-hour-a-day dream job. The ponytailed
Cummings worked as an activist aboard an anti-whaling vessel before
he showed up one day at the center's headquarters, where he initially
worked for no pay and made himself indispensable, focusing with
great success on stopping fishing fleets from laying waste to sea turtles and other imperiled marine life. Siegel worked as a raft guide in
Alaska before earning her law degree and coming to the center. She
won her first case in typical, frustrating fashion: In 2001, she represented the center and a group of other environmental organizations
in forcing the government to acknowledge that an Alaskan diving
seabird, Kittlitz's murrelet, was endangered—only to have it placed
on the warranted but precluded waiting list, where it remained eight
years later, its numbers dwindling.

Partly because of their frustration over such hollow victories,
Siegel and Cummings devised a strategy for using the Endangered
Species Act to take on the big environmental battle of the day, global
warming, then sold Kieran Suckling and Peter Galvin on the idea. At
first the center's cofounders were skeptical, but the attorneys laid out
their idea with irresistible logic: There is no greater threat to animal
species worldwide than climate change. Like the dinosaurs 65 million
years ago, most creatures will not be able to adapt to a rapidly changing climate. Warming oceans will kill coral, krill, and other creatures
that are vital parts of the marine ecosystem and food chain; this effect
will in turn drive larger fish, seabirds, and mammals toward extinction. Melting ice packs will be among the first visible sign, destroying
arctic habitats while also causing flooding elsewhere as sea levels rise,
destroying many of the habitats that the center had so painstakingly
fought to preserve. The same destruction of food chains and habitats
will eventually occur on dry land as well, as weather patterns are altered
and deserts march into temperate zones. Species already recognized
as threatened will be even more vulnerable. Their strategy would be
to carefully choose the right species, one for which climate change is a

very clear threat, instead of the more typical dangers—logging, construction, and pollution—found in most other endangered-species cases. Then everything else would fall into place: the mandates of the Endangered Species Act would compel the government to act against climate change.

As Siegel and Cummings saw it, the broad language of the Endangered Species Act did not limit the nature of the threat—the act had been written to be adaptable to a changing world, so it could respond as new threats emerged. The hard part, they said, would be the recovery plan. The prescription for the spotted owl was simple: this owl was dying out because its forest habitats were being logged. Prescription: stop cutting down the trees it needs to live. Global warming was different. It is *global.* There is no single logging operation, shopping mall construction project, housing development, or power plant emission that could be halted to save dying coral reefs from global warming. It's not as if the United States could suddenly stop pumping out 25 percent of the world's greenhouse gases just because the polar bear was placed on the endangered species list. That was a potential foil to their strategy: the *what next?* question. Opponents would undoubtedly raise that question trying to beat back the center.

But Siegel had a simple answer, an elegant solution that could have real impact on climate change, and that would simply rely on a process already in place: Section 7 of the Endangered Species Act, the consultation requirement that Suckling and Galvin used in their original spotted owl case. This section of the Endangered Species Act requires federal agencies to consult with the U.S. Fish and Wildlife Service whenever they are taking actions that could harm the protected habitats of animals on the endangered list. It is one of the act's most powerful and commonsensical provisions, because its purpose was to put all federal agencies on the same page instead of working at cross-purposes. The Transportation Department, for example, has to consult with fish and wildlife on highway projects in order to avoid building roads through protected habitats and crushing the endangered species living in them. The provision would work the same way for a species listed as endangered because of global warming: If the Energy Department was considering issuing a permit for a new power plant, under Section 7 it would have to consider requiring the owner to employ carbon-neutral means of electricity generation—solar, geothermal, nuclear, hydroelectric, or other clean technologies—instead of building yet another dirty coal plant that would spew millions of tons of greenhouse gases into the atmosphere, thereby contributing to global warming. Likewise, when the federal government purchases vehicles for its immense transportation fleet, whether postal service trucks or VIP sedans, or when it sets fuel efficiency standards for any car made or sold in America, it would have to favor electric cars and hybrids because of their much smaller carbon footprints. In short, as Siegel explained it, succeeding in listing a species endangered by global warming would compel the entire federal government and its panoply of projects, purchases, and permits to choose options that contribute the least to global warming and the extinctions it causes (and has already caused). This approach, which amounts to energy efficiency, conservation, and sustainability as national policy, ought to have been the way government operated for the past thirty years— there might not be any global warming if it had been and if other nations followed America's lead. But in the real world, it never seemed to happen without a legal mandate and a court order. Congress was too paralyzed and partisan to do it through legislation—even the most timid attempts at laws related to climate change had failed. No president has been willing to do it (though Jimmy Carter came close, and Barack Obama has promised more). But the Endangered Species Act could be the vehicle for that mandate, the attorneys believed.

Suckling and Galvin thought the plan was brilliant. It would shift the emphasis of the center toward the most pressing environmental challenge and threat of our time, they decided, yet still maintain the core mission of protecting biodiversity. As Cummings observed, what good is saving beautiful habitats if they end up under twenty feet of water? The only question, then, was what species to put at the center of what they hoped would be a landmark case. An earlier attempt—Cummings's successful listing of two types of coral, the organisms most visibly affected by global warming as rising ocean temperatures bleach and kill them—captured no one's attention or imagination. This is a constant problem for those working to save endangered species: Most of the creatures are not what you would call appealing or even moderately attractive. The center faces a continual battle for attention, public interest, and goodwill, and this is very challenging with many endangered species—imperiled insects, slugs, and unsexy, weedy plants. The threatened milk vetch is one of Suckling's favorites, but he may be the only person on the planet who gets sentimental about something called a vetch. These are species that are critical to the health of whole ecosystems and food chains; they can be repositories for important compounds, cures, and scientific discoveries (research on the milk vetch may aid in combating cancer and the effects of aging); and, from a moral or philosophical perspective, they are just as worthy of being saved and treasured as any other living species. But they are not lovable. It is all too easy—all too human—to focus on the downside of their protection: the fishermen who lose their livelihood when turtles are rescued from extinction; the off-roaders and hunters who are barred from their favorite haunts because of some burrowing rat or bristly shrub; the cattlemen who lose grazing permits and livelihoods, the local communities that may forfeit jobs, growth, and tax revenue when building projects are delayed, reduced, or blocked to save a toad that no one else but the center cares about. But this situation shifted when Siegel seized on *Ursus maritimus,* the world's largest bear, a ferocious yet beloved predator. The polar bear became the center's rock star.

Siegel formally petitioned to put the polar bear on the endangered list in February 2005. Her filing included extensive scientific documentation of the rapid melting of the arctic ice pack, without which the polar bear cannot survive, and evidence of bears starving and drowning as their habitat literally shrank beneath their paws. The petition was bolstered by a report from one of the most respected scientific organizations in the country, the U.S. Geological Survey, which estimated that two-thirds of the polar bear population of 25,000 animals[1] would perish by 2050, and that the species would go extinct before the end of the century because of global warming.

The Endangered Species Act has an inflexible series of deadlines; it requires (not suggests, not recommends, but requires) an initial finding by the U.S. Fish and Wildlife Service within ninety days. In this case, the government had ninety days to express a position on whether or not the polar bear was endangered—a finding that is purely scientific, or at least is supposed to be. If the initial determination is that a species is in jeopardy, this announcement must be followed by a proposed federal rule for listing the species as threatened (for long-term threats to survival) or endangered (for more imminent threats of extinction), which must be made public within one year after the petition is filed. And one year after that, after public comments are considered and a full scientific evaluation is completed, a final listing determination must be published and protection of the species must begin, along with preservation of its habitat and plans for its recovery. Two years, start to finish, are the absolute maximum the process is allowed to take before protections begin—there is no flexibility (except for the "warranted but precluded" loophole that the administration had been exploiting).

The first ninety-day deadline came and went after the filing of the polar bear petition, and the fish and wildlife service did nothing. This has been standard operating procedure since Bush entered the White House. From that moment on, the administration remained in constant violation of the law. After ten months, the center finally sued, and was joined by two other environmental organizations—the Natural Resources Defense Council and Greenpeace—which were partners of the center throughout the polar bear case. Then the center began a media blitz, with Siegel cast in the starring role. Her youthful face, her mane of long brown hair prematurely streaked with gray, and her tone of quiet, earnest outrage became a staple in the media whenever polar bears or global warming made the news—along with graphic, alarming satellite images of retreating arctic ice and heart-rending photographs of emaciated, starving polar bears. The public response to the center's petition—and to the administration's foot-dragging on protecting the polar bear—broke all records. More than 680,000 Americans filed comments with the fish and wildlife service on the need to protect the polar bear.

The government had no choice but to admit it was in violation of the Endangered Species Act's deadlines—it had no defense and made no apology—and settled the lawsuit by agreeing to speed things up. Finally, one year after the petition was filed—nine months late—the U.S. Fish and Wildlife Service made an initial determination that the polar bear deserved to be listed as "threatened." By then, the year-long deadline for step two in the process, issuing a proposed rule, was only one week away. The center and the other environmental groups agreed to extend the deadline from February 2006 to December 27, 2006, as part of settling their suit, reluctantly rewarding the govern-ment for breaking the law.

Secretary of the Interior Dirk Kempthorne called a press confer-ence on the day of this next deadline to announce that he agreed with the Center for Biological Diversity: The polar bear was threatened. Kempthorne didn't quite make the court-ordered deadline, as the actual listing proposal wasn't published for another two weeks. Even belatedly, however, this proposed listing represented a breakthrough: The Bush administration had conceded that the arctic sea ice was melting, destroying the polar bear's habitat—a first-ever admission by a member of Bush's cabinet that global warming was damaging a species and the environment.

But as she pored over the listing documents, Siegel soon saw the White House was continuing to play games: The proposed listing stated that the threat to the polar bear and the cause of its habitat loss could be neither identified nor regulated. Global warming, according to Kempthorne, was "beyond the scope" of the Endangered Species Act, as well as beyond the abilities of the fish and wildlife service, and so could not be included in the polar bear listing. "We don't have the expertise . . . to make those kinds of analyses." And so, he said, no such analysis was made.

None of the assertions Kempthorne made were true. Once again, the administration had resorted to suppression of science to alter the outcome of an endangered-species case. Siegel obtained the original report prepared by the scientists at the fish and wildlife service who had studied the polar bear, and found they had written an entire section in their report on how the federal government could protect the polar bear's habitat and stave off its extinction, a section headed, "Mechanisms to Regulate Climate Change." But Kempthorne censored that section and replaced it with a new one: "Mechanisms to Regulate Sea Ice Recession." Instead of the specific strategies for dealing with greenhouse gases and climate change that the scientists called for, Kempthorne's rewrite simply said there were no available methods to regulate the melting of sea ice. That was it. This was technically true but utterly misleading; only the *cause* of the melting—greenhouse gas emissions from cars, factories, and other human activities—can be regulated.

Protecting what is left of the bear's vanishing habitat was absolutely essential, Siegel had argued in her court filings, but Kempthorne offered a reason why there could be no such habitat protections for the polar bear: The Arctic was melting too fast and changing too much to be mapped accurately, much less protected. In other words, Siegel realized, the government might be compelled to list the polar bear as threatened, but Kempthorne seemed intent on doing everything possible to make the listing meaningless.

The government had a year—until January 9, 2008—to publish a final listing, at which time the polar bear would be entitled to full protection under the Endangered Species Act. That gave the center a year to increase public and congressional pressure for meaningful protections. Meanwhile, the polar bear's plight grew progressively worse. Researchers using satellite mapping discovered that during the summer of 2007 there was a record amount of sea ice melting in the arctic—1 million square miles of ice vanished. Projections that the polar bear's habitat could be largely gone by 2050 were revised—now it was looking as if this could happen by 2020, and that the rest of the world would not fare much better. As a climate scientist at NASA, Jay Zwally, put it, "The Arctic is often cited as the canary in the coal mine for climate warming. The canary has died. It is time to start getting out of the coal mines."

Polar bears, however, cannot leave their "coal mine"—evolution has left them supremely adapted to life on the ice pack, where they are efficient hunters, preying largely on seals and other marine life. But off the ice, they cannot hunt, mate, or shelter their young. They starve and die.

The January 2008 deadline arrived and passed with no listing emerging from the fish and wildlife service. The arctic was in an unprecedented meltdown, but the administration again broke the law, offering no excuses and no apology. Kassie Siegel had an explanation, and a week after the deadline expired, she had a national platform from which to offer it when she was called to testify before the House Select Committee on Energy Independence and Global Warming: The administration, she said, had delayed the listing so it could auction off huge leases for oil and gas drilling by major oil and energy companies along the Alaskan coast—in the polar bear's prime habitats. The auctions were scheduled for February. Because the center would offer evidence that the leases would pose a major potential threat to polar bears, they could never have been auctioned if the administration had listed the polar bear on time, Siegel testified. The department's own documents suggested a 50 percent chance that during the course of the leases there would be large oil spills that would devastate wildlife. "The Department of the Interior has illegally delayed protection of the polar bear at every turn and is now poised to auction off some of the species' most important habitat in the United States to the highest oil company bidder."

She was more blunt with reporters after the hearing: "Short of sending Dick Cheney to Alaska to personally club baby polar bears to death, there's not too much that the administration can do that is worse for polar bears than oil and gas development in their habitat."

Siegel sued again because of the delay, but it took until April to get a federal court to order the Bush administration to obey the law and issue a listing for the polar bear. The court gave the administration another month to issue it, and Siegel was called back to Capitol Hill again, this time before the U.S. Senate Committee on Environment and Public Works. By then the oil leases had been auctioned. "Polar bears are poised to become one of global warming's first victims," she testified, her slides of stranded, emaciated polar bears projected behind her. "The only thing keeping pace with the rapid melting of the sea ice is the breakneck speed with which the Department of the Interior is authorizing oil and gas development."[2]

One day before the court-imposed deadline, Dirk Kempthorne called another press conference to announce a final rule designating the polar bear as threatened because of global warming. It was a historic decision, humbling for the White House, which had fought this moment with all its considerable resources, and a huge victory for the Center for Biological Diversity. The polar bear would be protected, and the government had been compelled to admit that global warming was the problem. No matter how the administration tried to spin it, minimize it, or interfere with the protections for the polar bear, the listing gave the center and other conservation organizations a vast legal arsenal to use in court to force meaningful, perhaps sweeping, environmental reforms. Siegel knew that the listing was just a step in a very long process—but it was a huge step.

Kempthorne made it clear that he hated every minute of all this, and that he was making the announcement only because the law (or, more precisely, the center) compelled him to do so. He spent most of his press conference criticizing the Endangered Species Act for forcing his hand and for being too "inflexible." As Siegel expected, he rejected and mocked the idea of using the act to battle climate change, even though he and his department had reluctantly agreed that global warming would drive polar bears—as well as many penguins, seals, and other arctic and antarctic species—to extinction. He repeatedly lamented that the law barred him from considering "economic conditions" when deciding if the polar bear, or any species, is endangered. This was at best a bizarre criticism, because the entire rationale for the act was that before its adoption, deference to economic conditions had driven one species after another, such as sea turtles and bald eagles, to the brink of extinction.

Kempthorne had decided to build his own brand of flexibility into the polar bear listing, and he explained how he would hobble it from the outset by ordering that no new protections be offered beyond those already in place under the weaker Marine Mammal Protection Act—which he falsely characterized as stricter than the Endangered Species Act.[3] He vowed the listing would require no limits on oil and gas drilling in bear habitats, and that he would bar the listing from being "misused" to regulate greenhouse gases or to create "backdoor climate policy."[4] His position sounded reasonable to some people, but ultimately makes no sense legally, scientifically, or ethically. He had decided as a matter of law and science that global warming was killing polar bears, triggering a listing under the Endangered Species Act. Once an animal is listed, the act requires—not suggests, not recommends, but *requires*—the government to reduce the threat to polar bears, which means reducing global warming. Kempthorne simply did not have the legal authority to refuse to do his job. To do so was to make being an outlaw the official position of the Bush administration.[5]

In Joshua Tree, Kassie Siegel listened to a live Web broadcast of Kempthorne's press conference, savoring the bittersweet victory, if only briefly. When Kempthorne was done, Siegel helped draft a press release describing the listing as a "watershed event," and deriding the administration's position that the listing would change nothing. Then she sketched out the next lawsuit for the polar bear, one that would seek to make sure the listing means something big, even as the Bush administration began drafting new regulations to gut the Endangered Species Act and undermine Siegel's legal argument for strong polar bear protections. And then she turned her attention to the center's other big global warming case.

Tejon Ranch, one of the largest wild areas left in California and one of the twenty-five most important pockets of surviving biodiversity in the world—and the proposed location for one of the largest real estate developments in the nation—was back in the news. And the news, from the center's point of view, was not very good. A coalition of six environmental groups fighting the development had just splintered. Five of the groups, including the Sierra Club and the Audubon Society, had decided to accept the development after all, in exchange for the conservation of a substantial portion of the wilderness at Tejon—a deal that was being celebrated in the press and toasted by the governor. Only one of the groups attacked the compromise as a bad idea and vowed to fight on: the Center for Biological Diversity.

"This wilderness area is iconic," Siegel says. "It's California's heritage. There's no reason to put a new city there."

Tejón is Spanish for badger, a creature once plentiful near the ranch; legend holds that the first Spanish soldiers in the area, searching for deserters, found a dead badger at the mouth of a canyon instead, and the name stuck. The land had long been coveted for its fertility, beauty, and strategic location. It was occupied by the Yukuts and several other nations of Native Americans; was claimed by Mexico in the nineteenth century and carved up into four land-grant rancheros;

then finally became U.S. terrain after the Mexican War and California's admission to the union. Fort Tejon was established in 1854 by a storied California military man, explorer, road builder, and land baron, Edward Fitzgerald Beale. His many firsts included surveying for the transcontinental railroad, bringing news to Washington that gold had been discovered in California, and establishing an experimental U.S. Army Camel Corps on the ranch, with twenty-five of the desert animals imported from Egypt and Tunisia. Beale used his presidential appointments as head of the Bureau of Indian Affairs, Surveyor General of California and Nevada, and chief Indian negotiator for the U.S. Army to sequester the local tribes on reservations, then snap up the four *rancheros* at a fraction of their value to form the present-day boundaries of Tejon Ranch. He even bought some of the camels from the government when the brass decided they were not U.S. Cavalry material; he kept them on the ranch for years and is said to have used them to pull a surrey into town from time to time.

The Beale family owned the vast ranch for fifty-seven years. Edward's son, Truxtun Beale, sold it for $3 million in 1912 to a consortium of investors led by Harry Chandler, publisher of the *Los Angeles Times* and owner of the largest real estate empire in America at the time. The Chandler family's Times-Mirror Company converted the Tejon Ranch Company into a publicly traded corporation on the American Stock Exchange in the 1970s, then saw its stock prices quadruple to more than $400 a share in the early 1980s when the company first announced plans for massive development on the property. Drought, a bad economy, and a real estate downturn later stymied those plans and tanked the stock, and in the late 1980s and early 1990s the ranch owners positioned themselves as stewards of the land who were so wealthy they didn't have to worry about developing the property to make money. The owners were in it for the "long haul," the company president said at the time. In keeping with that philosophy, biologists from the U.S. Fish and Wildlife Service and the National Audubon Society were given frequent access to the ranch for research and wildlife management as part of the California condor recovery program, one of the most ambitious and successful programs ever attempted for staving off a species extinction. The last twenty-two condors in the wild were captured in 1987, bred in captivity, then "re-wilded" beginning in 1991—an expensive and painstakingly slow process that is still going on. By 2008, 146 condors were living free in California and Arizona. The company issued environmental reports in the early 1980s that recognized the importance of Tejon Ranch to the condors and the adverse impact that altering the landscape could have on them.

But a few years later, after Times-Mirrror sold its controlling interest in the ranch to several investor groups, a new management took over, and soon development was back on the table, bigger than ever, as the planning began for the current real estate proposal—a new city, a resort, golf courses, businesses, and industry in the midst of a pastoral ranch and sprawling wildlands. Cooperation with the condor recovery program was curtailed, and Tejon Ranch Company sued the government to keep the giant vultures from being re-wilded in their historic range near the ranch. The company sued again in 1996, demanding that the protections be lifted and that the nearly extinct condors be designated "experimental nonessential," a designation that would probably doom the species. The company then reached an agreement with the fish and wildlife service to put the lawsuit on hold if the federal government would give it permits to destroy condor habitat and even kill the birds during construction—an agreement the feds made even though it appears to violate state laws that provide absolute protection of the condors.

Siegel and the Center for Biological Diversity fought construction of an earlier industrial trucking and warehouse complex at the edge of Tejon Ranch abutting an already busy freeway, delaying the project for five years but ultimately losing the case. In 2005, when the housing and resort plans were announced, the center joined with a coalition of environmental groups and local activists to oppose the project and to propose taking further development off the table in favor of making Tejon Ranch a national or state park. In addition to the nearly extinct California condor, an estimated twenty other species protected by state and federal laws live at the ranch, as do another sixty rare types of plants and animals with no legal protections. Some of these creatures live nowhere else in the world. One-third of the oak tree species in California, including some of the oldest and largest oaks in the state, live on the ranch. This abundance and variety of life are why the region Tejon Ranch dominates, the California Floristic Province, is designated by the United Nations as a biological diversity hot spot—containing the last wildlife corridor linking the coastal, desert, and mountain regions of the state. Portions of the ranch have been designated as condor "critical habitat" under the Endangered Species Act, areas deemed vital to the species' continued survival. Yet the proposals for developing Tejon Ranch have called for building in the critical habitat and the ranch has sought a federal permit that allows condors to be disturbed, harassed, and even killed. This is euphemistically called an "incidental take permit." No presidential administration has issued more such endangered species take permits than President Bush's, with a long list that even includes blanket permissions to the oil industry allowing the killing of polar bears. A group of eleven condor experts who worked on the species recovery project oppose the development plan because of the threat it poses to condor species survival, but their input was not included as building plans were laid.

The Tejon project is not merely a disaster for endangered species, Siegel argues, but also exactly the sort of development that must stop if there is to be any hope of reversing disastrous climate change. Planting a city and a resort in the middle of an irreplaceable wilderness, with long commutes in every direction, during a climatic catastrophe and deep uncertainty over energy costs and supplies, is sheer madness, she says.

Fighting such "leapfrog" development—new housing projects

that are unconnected to existing towns and cities, and therefore have a much greater carbon footprint—is one of the new priorities for the center. And although California under Governor Arnold Schwarzenegger has led the nation in adopting new global warming legislation, a much older state law on the books, dating back to 1970 and signed by none other than Governor Ronald Reagan, has become the center's weapon of choice against greenhouse gas emissions and the urban sprawl that helps generate them. The law signed by Reagan, the California Environmental Quality Act, included the deceptively simple but sweeping requirement that local and state governments must examine *and reduce or eliminate* the negative environmental impacts of new development projects—from new cities to new shopping malls—before approving them. The goal was to address traditional water and air pollution—the mucking up of rivers and the smokestack smog and soot that were the bane and the primary environmental concern of the 1970s. Global warming and greenhouse gases were not on anyone's radar then and are not specifically mentioned in the act, but like the sweeping federal environmental laws that came a few years later, the language was designed to be inclusive of new environmental threats as they emerged, making them broad enough to address the greenhouse gas emissions of development projects and their impact on global warming. In practice, no one ever thought to apply the law in this way, until the center tried it out in 2006, in the first such lawsuit in the country.

In what Siegel promised would be the first of many such cases, the center went to court to stop a controversial leapfrog development called Black Bench, which would put nearly 1,500 homes in a wild desert area outside the city of Banning at the foot of the San Bernardino Mountains. The center accused the city of Banning, which had approved the remote Black Bench development, of failing to consider the increased greenhouse gas emissions it would cause, in violation of the California Environmental Quality Act. Citing wildfires, drought, and energy shortages as the initial consequences of global warming in California, Siegel argued that the city had two basic choices: cancel the development, or alter it to minimize greenhouse gas emissions by building in superefficient appliances, energy-efficient construction, passive and active solar energy systems, and requirements that the city be served by clean public transportation or that residents use zero-emission vehicles. Local and state governments ought to discourage "leapfrog" developments such as Black Bench that are located far from existing city neighborhoods and infrastructure, she argued, and favor development that is contiguous with existing urban areas. Keeping wilderness areas intact and concentrating development aids in the battle against global warming, whereas sprawl contributes to the problem because of longer commuting distances and the higher energy, water, and wastewater demands. The judge hearing the case agreed that the city had failed to properly consider the development's impact on the environment and overturned the city's approval of the project.

The center's strategy against Black Bench was so successful that the former governor and current state attorney general, Jerry Brown, immediately filed a similar suit against the entire fast-growing county of San Bernardino, forcing a settlement requiring the county to consider greenhouse gas emissions in every future development and construction project. Under threat of more suits from the center and the state, almost every other major jurisdiction in California is now doing the same—a sea change in how developers in California are required to handle the threat of global warming. Workshops began convening statewide to train city and county officials to "green" their urban planning, shorten commutes, and build in cleaner transit. Stopping a 1,500-home project like Black Bench will have no measurable effect on global warming, Siegel says, but how about ten such projects? Or a hundred? Or the multiple huge developments proposed for Tejon Ranch? No one project will make or break climate change—just as no one clean electric, hybrid, or hydrogen car will. It is the cumulative effect of millions of clean cars—or millions of houses—that can alter the path of global warming. "We have to start somewhere," is Siegel's litany.

On the other side of the issue, those who own Tejon Ranch, who have invested in the vision of a new city and resort complex set amid windswept hills and oak groves, see this as a simple question of property rights. They own the landscape. They are entitled to profit from that ownership. They say they have bent over backward to set aside significant land for a nature preserve and open space in their plans. They have invited outside environmental groups to oversee the conserved lands. What more must they do to satisfy the center? Most of the Tejon Ranch project is to be in semirural Kern County, where the county seat is rapidly growing Bakersfield, and civic leaders there have complained that the Center for Biological Diversity is going too far, and should not try to block development that promises jobs and revenue for the region. A columnist for the *Bakersfield Californian*, Marylee Shrider—in an article prominently reprinted on the Tejon Company's Web site—summed up these sentiments, deriding the center's "saber rattling" and "unwarranted sense of entitlement" as unwelcome environmental extremism. "They want, they want, they want," Shrider wrote of the center. " . . . Tejon Ranch Co. must develop, or not develop, the land according to their plan or it's off to court they'll go."

Early on, and despite its uncompromising reputation, the center had participated in discussions with the Tejon Ranch Company, joined by the Sierra Club, Audubon California, the Natural Resources Defense Council, the Planning and Conservation League, and the Endangered Habitats League. At the time, the Tejon Company had offered to place 100,000 acres of the 270,000-acre ranch in a conservation trust if the environmental groups would drop their opposition, but Peter Galvin, who personally represented the center in the negotiations, pronounced that insufficient. Much of the 100,000 acres was rocky, difficult high ground with little value for conservation and unusable for development—the ranch was giving up nothing, he said. But Galvin proposed an alternative.

The factions had gathered in the back room of a posh Italian restaurant in Pasadena—neutral territory, no one's turf. On one side of the big table sat ranch executives with the weathered, wise look of very prosperous cowboys, buoyed by their tablemates—their corporate allies and attorneys dressed as if for a day in court—confident, at ease, and seemingly in control. Across the water glasses and baskets of bread sat the loose-knit group of environmentalists, most of whom were hoping for a compromise, not a fight, and who seemed content to let Galvin play the "bad cop" in this drama. So he rose to make his soft-spoken pitch—a man who once spent weeks camped out watching a nesting pair of bald eagles, keeping a diary for the forest service of their every movement and meal day after day, hidden in a blind, cramped and cold and feeling utterly happy, privileged beyond measure. For him, the Tejon Ranch development plan was another example of an environmental holocaust, the sort of thing that had gotten him involved with conservation activism in the first place. He started by explaining that the best possible outcome would be to drop the plans to build a city of 70,000 called Centennial on pristine grasslands and a resort called Tejon Mountain Village abutting the condor foraging grounds and instead sell the land to the state or the feds to form a permanently protected preserve or park. Tejon Ranch Company would go down in history alongside the Rockefellers and Carnegies for its environmental generosity, its vision, and its commitment, he said. Generations would take joy in this beautiful wilderness and honor the decision. And of course, there were tax benefits.

Everyone in the room knew before another word was uttered that this idea was not going to fly. Merely selling mansions in the ten canyons where the condors once nested would bring in a profit of at least $350 million, and that was just a tiny piece of a very big puzzle. The investors in New York had not bought Tejon Ranch in order to play philanthropist—they had no interest in being eco barons. They wanted returns.

Then Galvin laid out his backup plan, the one everyone on the green side of the table would accept. He said he didn't like it, but here it was: Go ahead, build on 30,000 acres—the whole 70,000-person Centennial project, if you must. Just move it to a less sensitive area, away from the rare native grasslands currently slated to be bulldozed, and we won't sue. He had an alternative site already mapped out near the edge of the wildlands, and he argued that it would actually increase the profit potential because it would be less remote and so easier to build on. In return, he said, the ranch would have to cancel the resort complex and canyon mansions near the condor areas. The investors could still make a mint, and nature would fare far better.

Before the other side could do little more than bristle at that outrage, Galvin delivered the kicker: The amount of permanently preserved land had to be much more than the 100,000 acres already offered. It had to be 245,000 acres—about 90 percent of the entire property. Galvin said his biologists had determined that this was the absolute minimum needed to keep the various habitats and wildlife corridors intact.

"Nothing short of that will be acceptable," Galvin told the suddenly steely-faced ranch people. The alternative would be spending the next fifteen years in litigation.

The offer was rejected, as Galvin expected. Not long afterward, he and the center stopped participating in the meetings and negotiations, balking at the Tejon Ranch representatives' insistence that all talks be subject to a confidentiality agreement. What Galvin didn't expect was the deal that the Sierra Club, Audubon, and the other environmental groups struck with the ranch owners in May 2008, in which they made the extraordinary promise not to oppose any of the developments—even before detailed plans had been drawn up. In exchange, Tejon Ranch agreed to part of Galvin's proposal, offering up 90 percent of the land for conservation and open space as he had suggested. But this compromise left the vast development plans mostly unchanged, except for pulling back a bit from some of the ridges where the condors foraged. The agreement also allowed nearly one-third of the land that was offered for conservation to be developed, too, if the environmental groups failed to raise millions of dollars to buy the property at market rate within three years.

A media event was held at the ranch, featuring Arnold Schwarzenegger and one environmental leader after another praising the deal. Joel Reynolds, senior legal director for the Natural Resources Defense Council, called it "one of the great conservation achievements in California history . . . the Mount Everest of conservation in California . . . a once-in-a-lifetime achievement." Bill Corcoran of the Sierra Club called it "the ecological equivalent of the Louisiana Purchase." The *Los Angeles Times* proclaimed it a "landmark plan . . . that ends years of debate over the fate of an untrammeled tableau of mountains, wildflower fields, twisted oaks and Joshua Trees."

Oddly missing from the official celebration by the environmentalists and company officials was any mention of the fact that 26,000 homes as well as hotels, condos, golf courses, and an industrial center were still going to be plopped in the midst of the wilderness that was being celebrated. Nor was the fact that local activists for the Sierra Club, who actually lived near the area slated for development and adamantly opposed it, had been excluded from the process and the press conference. They were forced to resign from leadership roles in the local Sierra Club chapter so they could continue to voice opposition to the development. Now the Center for Biological Diversity is their last hope.

"This agreement is going to make it harder for us to win," Galvin says. "But that doesn't mean we won't win. . . . If we can't save a pristine piece of wilderness that the United Nations considers to be one of the twenty-five most biologically important on earth, what can we save?"

This is why the Center for Biological Diversity and its leaders are

eco barons. They do not flinch. They do not care if the top environmental organizations in the country think the Tejon compromise is the
best deal ever, because at the center, they see a compromise in name
only, in which only the environmentalists actually gave up something—
their legal and First Amendment rights—while the company gave up
nothing it really cared about. Where the news media see a triumph for
the environment, the center sees a disaster.[6] The people at the center
are willing to be hated, because they are certain they are right.

The battle over Tejon will last for years, and for the center, the
stakes are as high as in the case of the polar bears. If the center can
save the Tejon wildlands, if the project can be stopped or molded
into something environmentally sound, if consideration of extinction
and global warming can be made to trump money and sprawl here,
on California's last frontier—its *literal* last frontier—then Galvin sees
Tejon Ranch as the start of something big, something nationwide, a
seismic shift. It will mean America is no longer stuck on the old questions of how and why we should take action against global warming
and extinction, he says. We will have shifted to the questions of how
much and how fast we should act.

One set of questions leads to a world for our grandchildren in
which condors and polar bears still live in the wild, Galvin and Suckling and Siegel say.

The other set of questions relegates those species to life only in
history books. And they will be but the canaries in the coal mine.

Part Three

waiting for
thoreau

The Battle to Make America's Greatest Park

*Every creature is better alive than dead, men and moose and pine-trees,
and he who understands it aright will rather preserve its life than
destroy it. . . . It is the living spirit of the tree, not its spirit of turpentine,
with which I sympathize, and which heals my cuts. It is as immortal as I am,
and perchance will go to as high a heaven, there to tower above me still.*

—HENRY DAVID THOREAU, *THE MAINE WOODS*

a plum in the wood basket

The future of the last great wild forest east of the Rockies came striding through the sunlit fairgrounds of Unity, Maine, with dark hair flying and long beige skirt snapping in the autumn breeze. Roxanne Quimby stepped past the sheepdog demonstrations and the organic orchard, past the beehive exhibits and the folk art tents, past the solar panels and seventeen different models of composters, and then spied a shady spot on the grass outside the social and political action tent. She wound her way through the crowds with the practiced ease of someone who knew the Common Grounds County Fair well. She had spent many a weekend in years past hawking her wares at this annual gathering to promote organic farming—her homegrown honey, the beeswax candles, the lip balm in quaint small tins she painstakingly prepared on an old wood stove. She would work her small booth all day, then spend the night in her truck because she couldn't afford a

motel. Then it was up early the next morning to restock her display
with the homespun products of her tiny company, Burt's Bees.

But that was thirty years and many seventy-hour workweeks ago,
and today she had come to sell ideas, not products. Her company
wasn't small anymore. It had become a 60-million-dollar-a-year enterprise, the leader in natural personal care products such as sham-
poos, salves, and her trademark lip balm, featuring the funny little
pen-and-ink drawing of a sharp-nosed, bearded fellow in a railroad
cap: Burt, a melding of myth, man, and marketing that took off in a
way no one could have predicted. In the last few years she had cashed
in and turned her formidable and iconoclastic business skills—not to
mention a few hundred million dollars—in another direction, conservation. Her focus was to chart a possible future for the immense
Maine Woods in which spruce, pine, and icy blue lakes prowled by
moose and lynx and loon would trump the real estate investors' vision
of resorts, golf courses, and suburban homes on clear-cut lands.

In the not so good old days, Quimby likes to say, the forces of
progress and industry routinely crushed the tree huggers with barely
a glance. Now some sort of balance would have to be struck, a weighing of public good versus private gain for the vast and rugged part of
Maine called the "Unorganized Territories," America's last ancient
forest, 10 million acres of largely unoccupied woodlands, a wet, green
place on the map with more square miles than the entire state of Connecticut, a place where air and water are cleansed by nature by the
millions of tons. Here Quimby's conservation vision was up against
America's largest landowner, a timber and real estate company with
a reputation for rapaciousness and with roots stretching back to the
robber barons. The company was looking for windfall profits on tim-
berlands it had picked up at a bargain price—assuming its plans were
approved. Quimby was more interested in a long-term investment
in land and trees, focused not on profits but on her "mission"—her
"third career."

On paper, she seemed outmatched, an idea that made her smile, because the one real certainty in this contest is a matter of record: The woman who started her multimillion-dollar enterprise in a log cabin with no electricity, water, or phone has always thrived on being underestimated.[1]

About 14,000 years ago, when our ancestors hunted with flints and spears and a toothless old man might be all of thirty-two, the glaciers that covered half of North America began to retreat. The frozen tundra that would later be known as Maine (along with the rest of the northeastern United States and Canada) warmed as the last Ice Age waned, and life took hold in an explosion of green. Gradually, over thousands of years, the land was colonized by vast expanses of pine, fir, spruce, beech, and maple—towering trees that covered the region and stretched west as far as Minnesota, a forest for the ages. Deer, elk, moose, bear, wolves, lynx, eagles, hawks, and innumerable other birds made the forest their home. Thousands of lakes, rivers, and ponds teemed with trout and salmon and remain an enormous and famously pure freshwater supply to this day.

The worldwide retreat of the glaciers warmed the planet and stabilized shifting coastlines, paving the way for the first permanent human settlements and the birth of civilization in the Middle East about 11,000 years ago, when Jericho's first settlers arrived to build a new way of life. Yet even without the year-round ice, the lands of northern Maine remained beyond the reach of civilization, a vast, silent kingdom of impenetrable forest canopies and frigid, unforgiving winters. The region continues to resist humanity's dominance and quickly reclaims artificial intrusions; the rusting hulks of old steam engines can still be found in forest clearings on half-buried rail lines that the logging companies painstakingly built, then abandoned after a mere six years of backbreaking use.

The Penobscot, Abenaki, and Passamaquoddy peoples eventually came to hunt and canoe throughout much of the forest in Maine, but the northern reaches remained barely touched and uninhabited, perceived as mystical, powerful, a force to be reckoned with and respected. By the time Europeans established the first settlements in coastal Maine at the beginning of the seventeenth century, old-growth trees dominated the forest—towering, majestic, and hewn greedily by the settlers for buildings, forts, and ships. Giant white pines, straight and strong, were marked with the king's emblem and reserved for use as masts on royal sailing vessels. Huge swaths of forestland were burned to make room for farming, but not in the northern reaches of the Maine territory, which seemed as daunting to the transplanted Europeans as to the Native Americans before them.[2]

Most of Maine and all of the North Woods were publicly owned and part of Massachusetts when the American Revolution ended. But in an experiment in privatization unrivaled before or since, the government of Massachusetts, followed by the leaders of the new state of Maine, decided to auction off not just a portion of the lands, as other states had done, but virtually all the public lands and forests. This served the dual purposes of paying off war debt from the revolution and encouraging homesteaders to settle in the sparsely populated territory, with its immense natural resources.

The government divided northern Maine into townships. The term "township" summons images of frame houses and shops on Main Street, but in this case it was a legal fiction: mere lines on a map that carved up the forest and the surrounding lands into six-mile-square blocks. These were then sold off in lotteries, auctions, and tax sales, and in a few instances in the form of grants to veterans of the Revolutionary War who had been promised land for their service, though only a small number ever managed to collect it. The privatization process started in 1783 and continued steadily for the better part of a century, reflecting the state's early and enduring libertarian personality. Little if any thought was given to creating any public parkland or preserve. The Maine Land Office sold the last parcel of public land in 1878, and for many years less than 2 percent of the land of Maine was publicly owned (by 2008, the proportion was closer to 7 percent)—one of the lowest proportions of all the states. (By contrast, 15 percent of Pennsylvania is publicly owned lands, as is 44 percent of California, 68 percent of Idaho, and more than 90 percent of Alaska.)

In the midst of this vast sell-off of the public domain, the forests of northern Maine captured the attention and imagination of America's great philosopher, naturalist, and writer, Henry David Thoreau. His book *The Maine Woods*—published in 1864, two years after his death—documented his treks of the 1840s and 1850s through the wilderness on foot and canoe, most of them in the company of Penobscot guides, though sometimes he had just a compass for company. At his famous, pastoral Walden Pond in Concord, Massachusetts, he championed a simple life close to nature, but in the far wilder forests of Maine he honed what was then a new and revolutionary idea about wilderness—as something to be valued, experienced, and preserved rather than feared, harnessed, or exploited. In the Maine Woods, he found "primeval, untamed, and forever untamable, Nature." Man needs wild nature, Thoreau maintained, to balance and cleanse the artificial life of the city, the pretensions of society, the stale air of the street, the corruption of political discourse. Before Thoreau, there had been attempts in England, France, and colonial America to preserve forests as valuable resources, and deforestation had been linked to soil erosion, drought, and climate change as early as the 1760s, after ecological disasters in several island colonies. But the notion of preserving nature without regard to any benefits to mankind—other than spiritual ones—had never been advanced to a mass western audience before Thoreau.

Dismissed as an oddity by many of his contemporaries—in his day, it was the wide-open spaces of the West that captured America's imagination about wilderness, not the dense foliage of the East— Thoreau is now one of the most widely read American thinkers of his or any other time. An ardent abolitionist and an early champion of the idea of civil disobedience in the face of a lawless sovereign, Thoreau has been claimed as inspiration and hero by such diverse and important figures as Martin Luther King Jr., Mahatma Gandhi, John F. Kennedy Jr., Justice William O. Douglas, John Muir, and the legendary environmentalist David Brower, and he has been recognized as America's (and perhaps the world's) first ecologist. Thoreau, in turn, said that the wilds of Maine were one of his chief inspirations and touchstones, and he feared for their future at the hand of man. In one essay he lamented that man's mission seemed to be "to drive the forest all out of the country . . . as soon as possible." He urged the creation of a national preserve in the Maine Woods before that wilderness was lost entirely.

At least during his own life, Thoreau's idea did not catch hold in Maine. His vision of a nature preserve could not compete with the profitable business of harvesting trees, which was growing rapidly in his lifetime and would become an economic leviathan in the last half of the nineteenth century. The port at Bangor shipped more logs than any other in the world, and by mid-century there were 1,500 sawmills in the state. The nascent mass media rapidly became an essential part of modern life in America and the world, a by-product of the new technologies of mass-produced photographs in newsprint, industrial-grade high-speed printing presses, and typesetting machines. The demand for newsprint presented an enormous business opportunity for those who controlled the supplies of wood pulp and paper. An explosion of mill construction and logging operations began in the Maine Woods as the first timber barons accumulated one township after another, consolidated their holdings, and vied for supremacy. By the beginning of the twentieth century, Maine had become the world capital for logging, lumber, and wood pulp. Log drives filled the rivers and lakes with tidal waves of cut trees, a mesmerizing sight. The log drives also wreaked horrendous damage on the rivers and fish habitats, as some riverbeds were dredged and bulldozed to facilitate the log flows. New towns sprang up, and resorts for vacationing, hunting, and fishing opened at Moosehead Lake, a scenic wonder that was the gateway to timber country. The lakeside town of Greenville became a bustling place where the forest economy dominated all other activities. Whole communities were formed for logging and milling, company towns that existed to mine the forest.

In time the industry consolidated, until seven large paper companies dubbed the Seven Sisters[3]—among them International Paper, Scott Paper, and Georgia Pacific—bought up most of the 10 million acres of the Unorganized Territories, turning the Maine Woods into a vast tree farm. For nearly a century, these mammoth companies owned the wild third of the state, maintaining the forest's size, clearcutting the old growth but allowing the young trees to mature. The timber companies and workers called this seemingly endless supply their "wood basket"—a constant flow of timber, which meant a sustainable economy, even if the trees were meager in size and quality compared with the towering pines of old. Before the Seven Sisters, the privately held land had traditionally been open to all for hunting, fishing, camping, and public recreation, and the big timber companies allowed the tradition to continue. Maine has unusually broad public access laws, dating back to a 1641 law still on the books that grants public access to all ponds on private land. To this day, old-timers in Maine consider "no trespassing" signs in the forest to be in bad taste, and even when such signs are posted, their effectiveness is questionable. By continuing to allow virtually unfettered public access to their lands, so long as it did not interfere with the harvesting of trees, the Seven Sisters generated a great deal of goodwill. Generous leases were offered for hunting camps, and in more recent years there have been trails for snowmobiles and all-terrain vehicles, and outdoorsmen, hikers, and birders—pretty much everyone with an interest or passion in the Maine Woods—could continue to treat these private lands as if they were public. Many people came to look on this tradition as a birthright. A frequently expressed observation was, "Who needs parks when we've got the paper companies?"

Even with the logging roads and mill towns woven into the forest, and the public access granted to hunters and campers, most of the vast woods remained unsettled and filled with wildlife. Much of the forest is impassable in the winter snows and during the "mud season" that arrives with each spring thaw. The U.S. Census of 2000 found only twenty-seven people—five families—scraping out an existence in the 2,600-square-mile northernmost section of the woods, known as Northwest Aroostook. Those twenty-seven people occupied an area the size of Delaware. One hundred fifty years after his last journey there, Thoreau would have had no trouble recognizing the terrain.

But change was under way by then; the century of relative stability in the woods was about to end. The change started in the 1980s and accelerated in the 1990s, as the global wood and paper market shifted overseas to new centers thousands of miles from Maine. Enormous, inexpensive forestlands in South America and Asia, where wages were low, environmental regulations were minimal, and transportation was less daunting, replaced the Maine Woods as the leading global wood basket. One by one, the Seven Sisters began to close most of their mills and sell their holdings. Unemployment rose in the region as logging declined and the big companies pulled up stakes. Greenville—the town at the southern edge of the forest, the gateway to the scenic gem, Moosehead Lake—saw its jobless rate climb to twice the national average and its residents and businesses begin to move away.

Then, in 1998, the Plum Creek Company of Seattle, the largest private timberland owner in the nation, bought 1 million acres of the Maine Woods near Greenville, and promised a better future.

Plum Creek was a new kind of corporate entity in the Maine Woods—a real estate investment trust (REIT), rather than a traditional timber company. An aggressive buyer, seller, and developer of land that offered consistently high earnings to investors, Plum Creek delivered a 20 percent return in 2007 (five times what Standard and Poors delivered that year). Such bounty is not obtained by growing trees, despite Plum Creek's initial promise that it had no plans beyond the traditional timber harvest for its new land—a promise that would have been in keeping with the pattern followed by earlier land sales in the forest. Before Plum Creek, new owners continued traditional forestry and even sold off more than 1 million acres scattered through the Maine Woods to private conservation organizations.[4] Plum Creek's million acres were, however, irresistible to the company, encompassing the region's most spectacular scenery. Plum Creek had bought the shores, coves, and inlets of Moosehead Lake; the pristine Lilly Bay area; and a large slice of the Hundred-Mile Wilderness— the longest stretch of unsullied wildlands on the entire Appalachian Trail, where the 400-foot gorge of the Gulf Hagas, Maine's Grand Canyon, can be found. The land also provided important habitat for the endangered Canada lynx, a brown-gray big cat with distinctive black ear tufts, whose last refuge in the eastern United States was the Maine Woods.

As an officially designated "working forest," which is taxed at a very low rate to benefit timber companies, the 905,000 acres Plum Creek bought cost the company only $198 an acre. In working forests, Maine law permits the cutting of trees and related activities, but no other development. It did not take long for Plum Creek to figure out that if the land was rezoned and mansions, summer homes, and resorts were placed around Moosehead Lake and its spectacular views, those same acres would be worth more than $16,000 each. By 2002, the company decided to test this profit potential by requesting permission to build eighty-nine luxury homes twenty miles north of Greenville, in a remote, mostly wild area called First Roach Pond. Would Maine officials alter the law to accommodate Plum Creek, or would they impose limits and expensive environmental modifications to preserve the wilderness? Would buyers pay top dollar for such remote properties in an area not generally known outside Maine? And would the locals, whose average income couldn't come close to what was needed to buy one of the new properties on wildlands they had previously enjoyed free, express outrage or delight at the new economic activity?

The answers were everything Plum Creek could have hoped for and more. The plans for First Roach were quickly approved by the underfunded, understaffed, and weak agency in charge of zoning in the Unorganized Territories, the Land Use Regulation Commission (LURC). At the time, this was the single biggest development LURC had ever had to evaluate, and Plum Creek's lawyers, lobbyists, and planners outnumbered the LURC staff assigned to the case four to one. The public's perception was largely favorable—communities in desperate financial straits embraced the project and the prospect of new jobs and money entering the local economy. And finally, Plum Creek's offerings at First Roach were snapped up quickly, the bare lots selling for between $60,000 and $130,000 apiece, an enormous return on the investment.

Conservationists objected, as expected, but company officials said there was nothing to worry about. The development was small and there were no similar subdivision plans on the drawing board.

Less than two years later, Plum Creek announced plans to create in the Maine Woods the largest development in the history of Maine.

Beginning in the early 1990s, a New England environmental and advocacy group called Restore: The North Woods (with funding from, among others, Doug Tompkins) has advocated the creation of a large national park in the Maine Woods to preserve the still pristine landscape, and to forestall developments such as Plum Creek's.

In the vast system of America's national parks, Restore argues, there is nothing like the Maine Woods. Few existing parks could even come close to its scale: At 3.2 million acres, Restore's proposal would create one of the largest parks in the country, bigger than Yosemite and Yellowstone combined, and only slightly smaller than the largest park in the contiguous United States, Death Valley National Park.

It would protect some of the most significant terrain, scenery, habitats, and wildlife corridors in the northeastern United States; set aside areas for snowmobiling and other, traditional recreational uses; and still reserve two-thirds of the Maine Woods for commercial use.

"It's one of the last *big* places," Restore's director in Maine, Jym St. Pierre, likes to say. "There really is no other place this big, this wild. There's nowhere else we could be even be having this argument."

A soft-spoken outdoorsman and former state planning official, St. Pierre is passionate, reasoned, and convinced he has a vision that's right for Maine. He also may be one of the most patient men in the state, somehow maintaining his enthusiasm for a Maine Woods National Park despite a dauntingly long list of setbacks and disappointments.

"Everybody wins with Restore's plan," he says wistfully. Yet to date, despite fifteen years of lobbying and proselytizing, there is still no park. His organization is frequently reviled. He has been insulted, threatened, and derided as a carpetbagger trying to lock up the woods for radical environmentalists—notwithstanding the fact that he is a third-generation Mainer whose father and grandfather worked in the timber mills. "That's what happens when you propose a national park," he says resignedly. "It's a decades-long project."

Statewide polling suggests that a considerable majority of Mainers statewide like the idea of a Maine Woods National Park, and the cause has attracted some big-name supporters: Robert Redford, Harrison Ford, Jane Goodall, Walter Cronkite, Morgan Freeman, Don Henley, Holly Hunter, Meryl Streep. But the opposing forces are formidable, and no politician will embrace the park. The powerful hunting, snowmobiling, and gun lobbies are against it. So are leaders in Greenville and the other timber towns of the Maine Woods, who would rather see private development than a national park. Jym St. Pierre cites studies showing the proven economic benefits national parks have brought to surrounding communities elsewhere in the country, through tourism, jobs, and increased real estate values. But

Maine's old bias against public lands is still strong in those towns, where bumper stickers urging Restore to "Leave our Maine way of life alone!" have proliferated.

This is the longstanding problem big parks, both national and state, often face. Percival Baxter, a popular governor of Maine for two terms in the 1920s, was mocked as a crackpot, a socialist, and a closet homosexual after he proposed a large state park in the Maine Woods centered on the state's tallest peak, Mount Katahdin. The legislature refused to fund his idea. So he spent his personal fortune and more than thirty years of his life accumulating the land necessary for the park he envisioned, beginning with a 6,000-acre donation in 1930 and ending in 1962 with his twenty-eighth donation of land. By the time he finished, 200,000 acres had been preserved—more than 300 square miles—which Baxter insisted be kept "forever wild." Baxter State Park is now a much-loved destination and one of the most used and most popular state parks in the country—and home to some of the few pristine, preserved, old-growth forestlands left in the Maine Woods.

The same pattern of disdain followed by embrace has occurred throughout the nation. A proposed Grand Canyon National Park was denounced in editorials as a "fiendish and diabolical scheme" and its champion, President Theodore Roosevelt, as an "idiot." Now any attempt to decommission the park would be similarly attacked. In California, there were public protests against Redwood National Park by loggers and labor unions; and there were similar protests in the 1970s when President Jimmy Carter sought to expand the park's boundaries by adding the last remaining giant redwoods. Now a new generation of locals mourn the loss of ancient trees that a previous generation insisted be cut down in the name of progress. And even Wyoming's beloved Grand Teton National Park, which lies between Jackson Hole and Yellowstone National Park, was bitterly opposed by the local populace, for many of the same reasons Maine residents cite in opposing a Maine Woods National Park: loss of local control, restrictions on land use, an end to hunting (and in Wyoming's case, grazing) on public parklands, and a deep distrust of the federal government's intentions.

It took more than thirty years, from 1897 to 1929, for Congress to finally accept the idea of a Grand Teton park, and even then, the park that was authorized in a hard-fought compromise was cramped and incomplete. Only the mountain peaks themselves and their six adjacent lakes were included. Opponents of the park succeeded in excluding the surrounding lowland forest and range—the land that would make an accessible and usable public park. Opposition was so great, and local landowners were so suspicious of potential buyers, that the park's most important booster, John D. Rockefeller Jr., heir to the Standard Oil fortune and a dedicated conservationist, created a dummy corporation called the Snake River Land Company in 1927 to slowly buy up land near the Grand Tetons without revealing his involvement. When Rockefeller had accumulated 35,000 acres and his intention to donate it for a park became known, there were mass protests in Jackson Hole, Rockefeller was burned in effigy, and the Wyoming congressional delegation demanded that he be investigated for fraud. He was forced to testify before the U.S. Senate in 1933; and though he was later cleared of any wrongdoing, resistance in Wyoming and in Congress remained so great that it took him ten years more to persuade the U.S. government to accept his $1.4 million donation of land. President Franklin Delano Roosevelt finally accepted the donation in 1943 in the midst of war, combined it with adjacent public lands, and declared the creation of the new 221,000-acre Grand Teton National Monument. Establishing a national park requires an act of Congress, but under the Antiquities Act of 1906, a president can create a national monument by executive order alone.

FDR's unilateral decision outraged the Wyoming delegation to Congress, and legislation cleared both houses that would have overturned the monument designation and killed Grand Teton, but Roosevelt vetoed it.[5] It took until 1950 before Congress finally agreed to combine Roosevelt's monument with the existing park to make the current, highly popular Grand Teton National Park. It is now a beloved part of the local economy, attracting tourism and raising real estate values; and it is one of the ten most popular national parks, with 2.6 million visitors annually.

The late congressman and conservationist Morris K. Udall of Arizona, who sponsored legislation in the 1970s that over time doubled the amount of land in the national park system, observed in his autobiography, *Too Funny to Be President*, "I've been through legislation creating a dozen national parks, and there's always the same pattern. When you first propose a park, and you visit the area and present the case to the local people, they threaten to hang you. You go back in five years, and they think it's the greatest thing that ever happened."

Whether the Maine Woods National Park could follow that same pattern is unclear. St. Pierre certainly thinks it can. As he has often pointed out, at the $200 an acre working forest price, the land needed could be purchased for less than the price of one fighter jet or two days of the Iraq War (at the rate it reached in 2008, $10 billion a month). Americans spend twice as much on Christmas trees each year as would be needed to buy land for the park.

But Restore's plan for a park could remain viable only so long as the Maine Woods remained untouched by sprawling development. Plum Creek's plans for resorts and suburban development in the heart of the woods could mean the end of the park dream, and so Restore became its most vigorous opponent. This, in turn, hardened opposition to the park, as the long-suffering mill towns embraced Plum Creek as the improbable savior who would return prosperity to the forest. Yet it seemed clear that Plum Creek's plans, far more than Restore's proposal, called for transforming the face and character of the Maine Woods dramatically, altering landscapes that people and animals have used and depended on for generations, with uncertain benefits for anyone other than the company's investors. The Plum Creek proposal envisioned a sprawling development in pristine areas that included 975 home lots, additional space for condos and town homes, two resorts, three RV parks, and a variety of commercial and industrial areas to support this new community. The new housing—primarily luxury and vacation homes—would dwarf nearby Greenville with its 700 home lots in both size and real estate value. The buildings would not be concentrated in a single area but spread around Moosehead Lake. From the architect's vantage point, every hotel guest and homeowner would have a fabulous view. But from outside—from the vantage point of a canoeist, fisherman, hiker hoping for a wilderness experience—critics feared the view would become one of suburban sprawl, a place of buildings, cars, and Jet Skis, not wildlife and quiet. The project would bring traffic congestion, added air and water pollution, greenhouse gas emissions, and destruction of important habitats for rare and threatened species. Regulations governing land use in the region would have to be rewritten just to make the extensive construction legal—the law at the time Plum Creek bought the land made such a development impossible.

The promise of local prosperity, which residents were counting on, appeared less certain. There would be temporary construction work and service jobs at the resorts, and Greenville might see increased shopping dollars, an increase in school enrollment, and possibly more patients for the cash-strapped hospital. Sales tax revenues would not grow, however—the planned resorts would be built in the Unorganized Territories, so local governments in Greenville and the other mill towns would receive no tax income from spending at the Plum Creek properties. Tourism for the area might be harmed by the development, critics feared, because visitors to Moosehead come to see moose, not condos. They come for quiet, unspoiled nature, not the roar of motorboats that the resort could introduce.

But Plum Creek's strategy was flawless. First it put forward a "concept plan" for the region, before the Land Use Regulatory Commission could complete its own plan. LURC's planners were supposed to have written a concept plan for the area that balanced wilderness, forestry, jobs, and development far into the future—a plan, in essence, for what would and would not happen over the next thirty years in the Maine Woods. Towns, counties, and states adopt such plans all the time, to preserve the character of parkland, residential areas, historic sites, commercial districts, and industrial parks. These plans prevent a car dealership from opening on a quiet residential street or an amusement park from erecting a roller coaster next to the Lincoln Memorial. With no official plan in place, Plum Creek filled the void. Instead of having to conform its development to the wishes of the people of Maine and a comprehensive plan for all of the Maine Woods, Plum Creek submitted a plan describing exactly what it wanted for its own holdings, and demanded that the people of Maine accommodate it.

The company's next move was to win the support of the 50,000 members of the state's major sportsmen and snowmobile associations—powerful lobbies with considerable clout in the state capital and backed by the even more powerful lobbies of the gun and snowmobile industries. Plum Creek offered them continued access to favorite hunting grounds and snowmobile trails if the project went through; and this offer, coupled with the fact that support for Plum Creek meant taking a stand against Restore and its proposed park, cinched the deal. The leadership of Greenville and another major mill town, Millinocket, as well as other smaller towns in the area, supported Plum Creek for similar reasons.

Next, Plum Creek succeeded in dividing Maine's environmental community by offering to place about 400,000 of its acres in various conservation and working forest easements, and putting select environmental groups in charge of that land—if those groups agreed to back Plum Creek's development plans (and paid the company $35 million). The Appalachian Mountain Club, an environmental group whose focus is protecting the Appalachian Trail, could not resist the offer of a 28,000-acre corridor in the Hundred-Mile Wilderness Area. The Forest Society of Maine and the Nature Conservancy, meanwhile, were offered control over 270,000 acres of forestland east and west of Moosehead. This would make Plum Creek's proposal not only the largest development project in Maine's history, but also the largest conservation project in its history, twice the size of Baxter State Park. But it would happen only if Plum Creek's development project was approved. Critics called this blackmail and "greenwash," but Plum Creek had played its cards brilliantly—until the fine print of its offer and its environmental record began to be noticed.

Other environmental organizations not included in the deal soon found loopholes in the conservation proposal that would allow Plum Creek's continued industrial logging on the "protected" land, as well as the construction of cell phone towers, the stringing of power lines, the dumping of wastewater, road construction, commercial water extraction, gravel mining, wind farms, and the building of backcountry "huts" as large as 5,000 square feet and three stories tall. Meanwhile, Plum Creek received a record fine of $57,000 for breaking Maine's timber harvesting laws in 2006, and had to be ordered to stop destroying important deer wintering areas and causing a survival crisis for the animals. And then there was the matter of the threatened Canada lynx. No protected critical habitat was ever designated for this rare creature, twice the size of a house cat, until a successful lawsuit by the Washington-based Defenders of Wildlife forced the issue. In November 2005, as Plum Creek sought to advance its new development proposal, biologists at the U.S. Fish and Wildlife Service proposed designating most of the Maine Woods—more than 10,000 square miles, roughly one-third of the state's area—as part of the lynx's critical habitat. Over half of Plum Creek's land in the Maine Woods, much of it in the proposed development zone, would have been inside the protected habitat. Critical habitat designation on its lands would not necessarily stop the Plum Creek project, but it would bring the feds into the process, would limit the size and placement of developments, and would probably require leaving untouched several of the pristine areas frequented by lynx that Plum Creek considered essential parts of its proposed resort and housing complexes.

Company lawyers and lobbyists went to work at the highest levels, asserting that Plum Creek would suffer grave economic hardship if compelled to observe safeguards for an imperiled species. The company recruited the Democratic governor of Maine, John Baldacci, and both of the state's U.S. senators, the Republicans Olympia Snowe and Susan Collins, to intervene on its behalf. It also asked Secretary of the Interior Dirk Kempthorne to arrange an in-person meeting with senior officials of the fish and wildlife service so that Plum Creek's representatives could plead their case outside the usual public comment process. Staffers for the senators sat in on the meetings, indicating the clout Plum Creek wielded.

None other than Julie MacDonald, soon to be investigated for political interference in endangered-species cases, was given the task of dealing with Plum Creek. The case for preserving habitat was strong: MacDonald's own scientists reported that the Maine Woods critical habitat was essential for the lynx's survival. In recent years, Maine-based biologists noted a precipitous decline in the denning and birthing of lynx in the woods. And there had been a mysterious shooting of several of the animals living in the woods—a federal crime, punishable by six months in prison and a $25,000 fine, though no one was caught. MacDonald, after meeting three times with Plum Creek's representatives, overruled the science and sided with Plum Creek's view. The entire Maine Woods was removed from the lynx's critical habitat—a huge victory for the developer.[6]

But then the MacDonald scandal erupted, as a result of the work of the Center for Biological Diversity, and the lynx decision became one of seven cases in which manipulation of the underlying science was so clear that the head of the fish and wildlife service reluctantly conceded they had become a "blemish on the scientific integrity" of the Interior Department. The question of critical habitat for the lynx was reopened. In February 2008, the agency once again proposed designating a large protected habitat for the lynx that would encompass almost all of the Maine Woods, nearly one-third of the entire state, including portions of the land around Moosehead Lake that Plum Creek wanted to develop. At Restore, Jym St. Pierre was ebullient. "Blocking habitat protection . . . was no way to demonstrate a commitment to protecting wildlife and the environment," he testified during a final round of public hearings before LURC. "How can the company be trusted on its Moosehead plan?"

But the controversy over the lynx and the proposal to extend its habitat did not provide a path to resolving the debate over the future of the Maine Woods—it only polarized the factions further. As record numbers of Mainers submitted comments on Plum Creek's proposal during the public hearing process, it became clear that none of the options on the table had generated much enthusiasm—not St. Pierre's cherished proposal for a national park, nor Plum Creek's proposed mega-resort complex. Both ideas had large, vocal groups of opponents. It seemed the perfect time to propose a third way.

This is when Roxanne Quimby made her move.

bees and trees

Roxanne Quimby's journey to the Maine Woods began when she fled the raucous counterculture bustle of San Francisco after earning her art degree, embarking with her boyfriend on a cross-country drive in search of a simpler life. Helen and Scott Nearing's 1954 "back to the land" manifesto *Living the Good Life: How to Live Simply and Sanely in a Troubled World,* had inspired her to seek out the agrarian lifestyle that they recommended and that Thoreau had sought before them: finding joy in quiet, in plainness, in work with your hands, in nature. It was 1975 and she was twenty-five, a child of the 1960s, disillusioned by Vietnam, by assassinations, and by Watergate. She was ready to drop off the grid and to see if, through getting close to nature, she might find for herself a different take on the vision that had captivated her European immigrant parents: America as a promised land. She would find it, too, though in an unexpected place and manner, as she became an even more unlikely CEO than the mountain man Doug Tompkins: astrologer, mystic, hippie, waitress, single mom, Horatio Alger character, magnate, idol of teenage girls everywhere, eco baron.

The journey started when Quimby and George St. Clair loaded up the requisite Volkswagen microbus and headed to Vermont, looking to buy a homestead. They soon discovered their $3,000 nest egg wasn't enough to buy much of anything in the Green Mountain State, not even bare land. "Try Maine," someone suggested.

On the edge of the Maine Woods, in rural Piscataquis County outside the mill town of Guilford, they found thirty forested acres they could afford, cool and fragrant with pine needles. They set about clearing a portion of the land and using the logs to build their own cabin, one big room with a loft for sleeping. There was no electricity, no running water, no telephone, just an old wood stove and some kerosene lamps. They hauled water from a cold spring on their land. Nothing was wasted: Water from washing the dishes would be used to wash the floor. In winter, they'd cover the spring to insulate it from the subfreezing temperatures, but eventually they'd have to fetch water with an ax. In the warm months they planted and tended a vegetable garden. When the VW died, they walked the two miles to Guilford. During the muddy spring, they wore high black rubber boots. During the snow season, they wore snowshoes. Sometimes they needed snowshoes just to get to the outhouse.

You can live a good life without much, Quimby decided, with the added benefit that a lack of resources makes you resourceful. It was idyllic in many ways, but hard work, too, especially in winter, when the cabin seemed very, very small, and the idea she had of living off the land, painting, and selling her work for what money she needed didn't seem to be working out very well. She took jobs as a waitress instead, but because of her innate rebelliousness and her desire for independence—"I do have problems with authority," she says mildly—her employment did not always end well. Showing up one morning to wait tables at the local diner in a conservative mill town with her head shaved was, in retrospect, probably not the best path to job security. She began buying and selling items at flea markets and yard sales instead.

Roxanne and George married, and in 1978 the twins Hannah and Lucas were born. Five years later, the couple divorced. The cabin was confining, and it turned out that their personalities were unsuited for the long haul. George, who would become a social worker, was steady and predictable; Roxanne was impulsive, whimsical, always ready for a change. As she would later explain it, her mercurial nature worried him; his constancy bored her. The combination covered all the bases in parenting, Quimby recalls, but didn't make for a successful marriage. So they agreed to joint custody of the twins, switching every other week, and she found some acreage in the area and moved into another cabin. The twins would be thirteen before she finally installed electricity and a phone.

One day in 1984, a yellow Datsun pickup truck pulled over for Quimby as she hitchhiked to Guilford. The driver was Burt Shavitz, a former photojournalist, now a beekeeper, who had also heard the call to return to the land. He lived in a converted eight-by-eight-foot turkey coop, and he kept chickens and thirty beehives. On weekends he sold honey out of his pickup truck; the honey was packaged in old pickle and mayonnaise jars with the original labels still on them, big and unappealing. He made about as much money as Quimby did at the flea markets—which is to say, not very much. Quimby suggested they might be able to put her artistic talents to work on better packaging and labels, and in turn he could teach her about bees so she could help with the hives. And so they forged a partnership, and for a time became romantically involved as well, although the work ended up far outlasting the romance. She was thirty-four and he was forty-nine.

Growing up—mostly in Lexington, Massachusetts, but in other towns, too—Roxanne and her sisters received no allowance. Instead, they were encouraged by her father, who worked in sales and had to move the family from time to time, to find ways to earn spending money on their own; whatever they saved, he promised to match dollar for dollar in a college fund. The Quimby girls were always baking cookies and making lemonade to sell in the neighborhood. By age twelve, Roxanne had also learned to make soap and stuffed animals to sell. She virtually stopped being a consumer when she moved to Maine, but she had a flair for what real American consumers like. She found some small plastic jars shaped like bears and beehives, then made her own stylized pen-and-ink drawings on yellow labels, featuring an old-fashioned beehive—the familiar dome structure made of concentric rings of woven straw called a skep. Shavitz had stenciled "Burt's Bees" on his hives. Quimby found this hilarious—she wondered, Could you really *own* a bee?—but she decided to put the name on the label. It was catchy and homey and people liked to read it and say it aloud. Sales of honey picked up.

Shavitz had stored up years' worth of beeswax in his honey house, hundreds of pounds of it. "Maybe you can make candles," he suggested. So she taught herself to work with beeswax and brought the results to the annual Christmas crafts fair at the local junior high school: candles for three dollars a pair, made from the all-natural wax of Burt's Bees. They made $200 in a day, which was good money for them back then, and they decided to make a business on the craft fair circuit. The initial investment was $400: a supply of wicks, a dipping tank for the wax, a couple of grosses of honey jars. Roxanne found an old book of nineteenth-century recipes for using beeswax, and soon she was cooking up shoe polish and furniture polish—neither became a big seller—and cute little tins of all-natural lip balm, with her inked drawing of Burt in his railroad cap, bushy-haired and big-bearded. The balm sold, and sold well—Quimby could barely make enough of it. She could see that the packaging was what people reacted to as much as anything else, because she watched the shoppers as an anthropologist might study a local ritual. Burt, unmistakably, had become a brand. He was her own Mr. Clean, her Michelin tire man, her Pillsbury Doughboy. He was the alternative lifestyle guy,

homespun, natural, and trustworthy, a safe little piece of the counterculture that also happens to make your chapped lips feel good. "We
could actually make some money at this," she told Shavitz, and even
then, when the business was little more than some old buckets of wax
and an idea, the woman who had thought she was "above" thinking
about money, who had rejected consumer America for a log cabin in
the Maine Woods, suddenly started thinking very differently about
the value of things, and the power of a few bucks in the bank.

They soon realized that Shavitz's turkey coop and Quimby's
cabin kitchen would no longer suffice as places of business. They
needed something big and cheap as a base of operations. A friend
owned an old, disused one-room schoolhouse in Guilford, which he
offered them for the unbeatable rent of $150 a year—the amount of
his annual fire insurance premium. It had no electricity and was missing glass in the windows, but the addition of a gas stove, some kerosene lamps, and a generous amount of cardboard made it a fine first
headquarters for Burt's Bees. Quimby and Shavitz began making the
rounds of craft shows and fairs around New England in 1985, renting
a booth for $200 and hoping their sales would cover the expense. It
entailed hard work and long hours for a modest living, but Burt's Bees
was tapping into a new market—natural personal care items. There
was demand, but little supply. After five years of hardscrabble work
and continual experimentation with new products—hand lotion and
soap were among the more successful—a lucky break came their way
in 1989. A Maine company dropped out of the New York Gift Show,
which is frequented by buyers from trendy New York City shops
and boutiques. Burt's Bees was offered the open spot, something that
they normally would have had to wait four years to get. A respectable stream of orders began to come in afterward. Later in 1989, the
historic, upscale Manhattan women's shop Henri Bendel, famous
for introducing top designers (including Coco Chanel) to American
shoppers, called and asked Quimby to come by to make a presentation the next day. She appeared in her boots and down parka and

delivered her spiel. Henri Bendel snapped up the whole Burt's Bees line and displayed the products prominently at the front of the store, and almost immediately the stream of orders became a flood. The door was opened. Suddenly Martha Stewart was using the obscure natural care products from Maine with Burt's face plastered on the label to make Mother's Day gift baskets on *Good Morning America*, announcing, "I love Burt's Bees!"

A short time later, Shavitz and Quimby hired their first employee. By the end of 1990, they had forty-four workers and had set up a factory in Guilford's former bowling alley. Most of the employees were women, many of whom had been receiving unemployment benefits or other government aid in the difficult economy of northern Maine, but Quimby found she was pretty good at cherry-picking raw talent. She didn't care about academic credentials, which seemed irrelevant, or technical experience, which could be learned. What mattered was how someone performed on the factory floor. What mattered were the work ethic, habits, attitude, how people followed, and how they led. "I've had people with no formal education beyond sixth grade outperform MBAs from Duke University," she says with pride. She and Shavitz gradually added a variety of products, including handmade dog biscuits and baby clothes made with organically grown cotton. Every product was touted as earth-friendly with no less than 90 percent natural ingredients. Most products had a considerably higher percentage (with the exact proportion on each product's label: Radiance Lip Shimmer, 100 percent natural; Baby Bee Skin Crème, 99.71 percent natural; Burt's More Moisture Raspberry and Brazil Nut Shampoo, 98.12 percent natural).

The business began growing 40 to 50 percent a year and by 1993, Burt's Bees was incorporated (a two-thirds share went to Quimby, one-third to Shavitz) and had annual revenues of $3 million. And it seemed to be outgrowing Guilford. There were several problems: The little Maine town was remote, and shipping costs were through the roof, sometimes exceeding the value of the product being shipped.

Then there were Maine's high unemployment taxes and workers' compensation costs. Quimby embraced numerous liberal causes and environmentally friendly practices, even when they might hurt short-term profits, but she had also evolved into a hard-nosed business-woman. She began to think there might be a better place for Burt's Bees to grow, where the cost of doing business might be lower.

The tipping point in favor of moving finally came out of Quimby's sense of outrage at government when an employee claimed an on-the-job injury, and Quimby subsequently learned that the same person had already received workers' compensation settlements from two previous employers for exactly the same injury. Not only that, but the employee had been arrested for breaking into campsites and stealing people's gear and possessions—while supposedly recovering from her injury. Quimby reported the former worker as a possible fraud, and a serial fraud at that. The Maine workers' compensation officials, Quimby recalls, informed her that they would not con-sider the employee's past actions or allegations about her off-the-job conduct—only her time at Burt's Bees was deemed relevant. Quimby, dumbfounded, predicted that with this standard of review, the worker would soon be driving around town in a new pickup truck purchased with a nice fat settlement, and further suggested that such an out-come would send the wrong message to the rest of the community: that you could make a good living beating the system and abusing employers. A few months later, Quimby saw the former worker with a big smile and a big new pickup truck. Burt's Bees started shopping for a new location the same day.

Quimby contacted the state economic development offices in North Carolina, Florida, and Maine—the first two because of their business-friendly reputation, and the third to be fair, to see if there might be some incentives available to relocate within the state. Both southern states overnighted full packets of material to Quimby, in-cluding a CD-ROM from North Carolina that allowed her to calculate her payroll taxes and workers' compensation costs, which compared very favorably with what she had to pay in Maine. The office in Maine never called her back. So Quimby and Shavitz settled on Durham, North Carolina—the cost of doing business would be lower, shipping issues would be resolved because Durham lies at a confluence of east-west interstate highways, and there was a ready supply of well-educated workers. Hannah and Lucas were thirteen by then and in boarding school, so the move would not disrupt their lives.

As they were packing to leave—with the departure set for the next day—the governor of Maine called. He had read an article in *Forbes* magazine profiling Burt's Bees, and it had mentioned that the company was moving because of the difficulties of doing business in Maine. The governor wanted to know if they could talk about finding a way to keep Burt's Bees in the state. Quimby told him sorry, no. "We've just finished loading up," she told him. "We're literally leaving tomorrow."

The move allowed the business to grow and prosper, but it was the end of the line for Burt Shavitz, and he returned to Maine and his turkey coop. Eventually, Quimby bought out his share of the company for the price of a $130,000 home in Maine, which he promptly sold in favor of the old coop, though he did add on some extra space. (Given the windfall she would later reap from selling the company, Quimby felt Shavitz's share was an embarrassment, and she later agreed to pay him another $4 million, though he has always said that his joy in life came from living off the land, not from money.) Shavitz continued to be paid annually for the use of his image and name, and to work as a kind of goodwill ambassador for the business. Burt became an essential part of the company mythology, his true history merged with an embellished tale of backwoods wit, gumption, and devotion to a natural way of living. Shavitz would visit stores and meet customers; the teenage girls who are an important constituency for Burt's Bees products would line up for his autograph, and the legendary Burt would hold court as if he were a retired member of the Grateful Dead.

In North Carolina, calling the shots on her own as CEO and creative director, Quimby modernized the company. The handmade items—the less than successful dog biscuits, the baby clothes, and even the signature beeswax candles—were jettisoned and the factory was automated. All this improved the bottom line, though the workforce grew in tandem with the company's growth. Quimby fashioned the company as a sustainable, green enterprise that used recycled materials and minimalist packaging—and that deplored waste. The lessons of her resource-starved resourcefulness in the Maine Woods informed her business practices, from assembly-line methods to the campaign she launched to have customers mail their empty tubes of balm back to the company for recycling.[1] In the early 1990s, this made Burt's Bees a trendsetter, and the green policies were both an end in themselves and a way to strengthen the brand, setting it apart from its competitors. In company literature and catalogs, Quimby explained that Burt's Bees was "setting the Natural Standard," that it was an "Earth Friendly" company, and that she had devised a way of doing business she called "The Greater Good Business Model." She said this made social responsibility and environmentally friendly practices an essential part of the company's DNA, not an afterthought, a "greenwash," or a reluctant sop to environmentalists. Pursuing The Greater Good meant writing a business plan that included such terms as "human rights," "animal rights," "sustainability," "well-being," and "fair trade." Burt's Bees deplored animal testing of products, artificial ingredients, and toxic substances. It rewarded employees and vendors who suggested greener practices. The company resisted launching its signature product, lip balm, in a plastic tube for years, until it persuaded a plastics manufacturer to use recycled plastic. There was an unintended ripple effect: As a result of finally bowing to Quimby's wishes, that manufacturer discovered a new, environmentally beneficial—and highly profitable—business model, and has become an industrywide leader in making recycled plastics.

The Greater Good model wasn't perfect—some compromises had

to be made. Gone were the days of locally grown beeswax from Burt's thirty hives, which had been about as environmentally friendly as anything gets. A company selling millions of dollars worth of beeswax and almond oil lip balm needed a huge supply of the stuff. The essential ingredient had to be imported from Ethiopia, a world leader in beeswax production using traditional hives (only China and Mexico produce more)—so each tin or tube of balm had a carbon footprint from global shipping. The company tries to make itself "carbon neutral" by purchasing carbon offsets every year, underwriting renewable energy projects that cut enough greenhouse gas to balance out the company's emissions.[2]

Quimby kept her desk in the art department rather than in an executive suite, and any employee could seek her out for a chat. She ate lunch with the other workers. She had wealth but still drove an old car and lived simply—not to promote the image but because that was what she wanted and liked. For similar reasons, Burt's Bees never advertised or conducted focus groups, although it religiously protected its brand identity and nurtured Quimby's counterintuitive insight that the scruffy but trusted image of old bearded Burt, the opposite of the glamorous models used to promote other beauty products, had made Burt's Bees a leader of the natural personal care category of products. She had found a way to bottle trust. It made her a millionaire, and being a millionaire in turn gave her the opportunity to begin to conserve land in the Maine Woods—what she calls, "My mission."

It would be easy to write off this success as the happy product of a stab of inspiration and a lucky break or two—and certainly both principles apply to Burt's Bees. But the bulk of the success of the company, as Quimby sees it, was built on many years spent in craft show obscurity, making long hauls to fairs in an old van with a blue shag interior, spending long nights squinting over the stinging vapors from cauldrons in a schoolhouse with cardboard for windowpanes, slim customer lists, slimmer bank accounts, and much trial and error. At one point she and the twins had to live in a tent at a campground because she couldn't make ends meet. She is rueful when she describes the hard times, but proud, too, and it is clear she enjoyed those years. She is a practical businesswoman, yet she remains a child of the 1960s, the queen bee of beauty products who eschews makeup for herself, a woman comfortable in her own skin, with a broad, infectious laugh and thick, expressive eyebrows. She is a master of artful plain talk in the classic Down East Maine tradition: "I have a very clear memory," she says drily, "of what it took to make the millions that I spent on that land. It wasn't fund-raising from wealthy donors. I wasn't taxing my constituents. It was my cash on the barrelhead."

By 2003, the barrelhead was impressive. Burt's Bees had 250 employees and $59 million in sales. Projected growth for 2004 was 40 to 50 percent. Burt's Bees' prospects could not have been brighter. .

Yet Quimby felt ready for a change. The old aphorism was true, she realized: "Business is just one damn thing after another." Two decades of Burt's Bees had begun to feel like a treadmill. Her next stage of life—the lure of conservation—beckoned. She had turned away offers for the company in the past, but now she was ready. The search for a buyer Quimby deemed suitable took a year; there were forty suitors, winnowed to a dozen, then six, and finally one: AEA Investors, a Wall Street private equity firm that had been founded by the Rockefeller, Mellon, and Harriman family interests. She stayed on as CEO one more year after the deal closed in October 2003, remained on the board of directors, and retained a 20 percent interest in the company. And she walked away with $179 million.

your land is my land

Even before the sale of Burt's Bees, Roxanne Quimby had returned her heart and her money to Maine. It was in her blood, it was where she wanted to live. She found a 1,200-square-foot home in the coastal town of Winter Harbor, near Maine's only national park, Acadia, another conservation project that had received vital financing and land from the Rockefeller family.

She also bought a place in Palm Beach, Florida, thinking that it would serve as a warm getaway during Maine's harsh winters, but she felt out of place there. The conspicuous consumption and public displays of wealth that are part of the fabric of Palm Beach, the Jaguars and Mercedeses her neighbors seemed to covet, are alien to her. She drives a Toyota Prius in Maine, and one of her best friends runs the local recycling drop for bottles and cans, and sells used furniture next door to it—exactly the sort of little homemade business Roxanne Quimby survived on for many years, the sort of business Burt's Bees once was.

A few years before the buyout, Quimby started casting about for a mission beyond Burt's Bees, something to invest in—her money, her time, her passion. She had been consumed by the business for so long, it had only just dawned on her: I've got all this money now, more than I need to live, more than I need to sustain the business. What's the best way to conserve that wealth? What will retain its value through changes of governments and politics and shifting economics? What endures? And when she thought in those terms, the answer was obvious: land. Buying pieces of the Maine Woods would be like investing in a long-term bond, but the forest would be nicer to look at, and better for the world. And since she had no intention of logging on her land, or hunting on it, or developing it, the land would just grow in value, slowly and surely. She figured that big trees are worth more than little trees, whether it's the oaks or maples in your own front yard, or a forest full of them. Much of the Maine Woods had been cut down and replanted over and over, every thirty to forty years for two centuries. Quimby knew those woods desperately needed some big old trees. Her land would be a haven for them, a wild sanctuary.

The idea appealed to her spiritual beliefs. Quimby is not a churchgoer, nor is she conventionally religious. She prefers to immerse herself in "alternative spirituality"—she's an astrologer, a tarot card reader, and a regular client of psychics. She meditates daily, at least an hour of yogic deep breathing, longer if she's "having issues." Whatever is worrying you before the meditation begins, she says, is less of a problem after. She has read a great deal of "soul journey" literature, and is particularly taken with the idea that animals show themselves to people for a specific reason, to offer up messages and clues. "I find that interesting," she says. "We ignore all the species but ourselves. We think we're so important, that every species is subordinate to us. How arrogant! We live without thankfulness, without humility." She is convinced that, absent some very meaningful and immediate reforms, the universe—through global warming or some other cataclysmic event in the near future—"is going to put humanity in its place in a very humbling way." She is a believer in karma, the Hindu idea that we bring on our own rewards or punishments by our actions. "Humanity's karmic debt is off the charts," she says. "It's not going to be pretty."

The second part of Quimby's quest for an enduring investment came to her in the form of some promotional materials from Restore describing its proposal for a 3.2-million-acre Maine Woods National Park. The vision of a gigantic national park in the woods she knew so well was an epiphany for her. It's exactly the sort of project she felt must be undertaken if there was to be any hope of staving off her predicted humbling of humanity. And the proposal brought back childhood memories of road trips with her parents to Redwood National Park, the Grand Tetons, Grand Canyon, and—back east—all the national battlefield parks of the Revolution and the Civil War. She had a national parks passport, and each time she visited a park, she'd get a different stamp. She had almost forgotten how much she had enjoyed those traipses. National parks, in Quimby's view, showed the best of America, the ultimate democratic experience—ten bucks' admission bought a day of entertainment for the whole family, rich or poor, open to all.

So in 2000 she began making land purchases in the woods: 5,000 acres here, 8,000 there. She found a real estate agent in Bangor who knew the forestlands well, and he called her regularly with early tips about prime land, suitable for a wildernesss sanctuary, about to come on the market. She wanted to focus especially on the forest, wetlands, and mountainous terrain surrounding Baxter State Park, areas that the former governor had always wanted to acquire but never could—land that would enhance and buffer the park, and also form the starting point of a new national park that Quimby wanted to help endow. Baxter, along with Thoreau, had become her inspiration; she hoped, though she was reluctant to articulate the idea, that she might someday be remembered, like Baxter, as a champion of preserving the Maine Woods. Quimby formed two charitable foundations to carry out the work—the Quimby Family Foundation, which makes grants in the arts and environmental projects; and Elliotsville Plantation, named for her first land purchase, which administers her conservation lands and employs a full-time ecologist, Bart DeWolf, formerly of Restore.

Quimby made her initial purchases in relative obscurity, amid a mad scramble of real estate transactions in the Maine Woods as land changed hands more in a year's time than had happened in the previous century. But then in May 2003, Quimby joined the board of Restore as part of its campaign "Americans for a Maine Woods National Park," in which 100 celebrities, artists, businesspeople, and others signed on to support the effort. A press conference on the initiative was covered by the media throughout Maine, and Quimby was quoted in her role as the campaign's cochair. She made it clear that her land purchases would be part of the national park effort. Her days of purchasing forestland in relative anonymity were over.

She next bought a 24,000-acre township on the northeast border of Baxter State Park—an area prized by hunters, snowmobilers, and loggers. Opponents of the park pounced, complaining about anticipated new restrictions on the land by the conservation-minded Quimby—even before she had closed the deal. They called and wrote to other landholders in the forest who were considering selling, warning them not to do business with Quimby and her "cosmetics empire" if they cared at all about the Maine way of life.

As it happened, another 47,000 acres next to Baxter State Park were auctioned by the same seller at the same time, and were bought by two logger-developers, one of them a voracious "forest liquidator"—a company that clear-cuts a forest and then sells it for housing development. This land had been sought by the Maine Department of Conservation; Percival Baxter had tried in vain to buy it many years before for his state park. It is spectacular terrain that includes a last remnant of old-growth and mature forest trees; a pristine remote stream, the Wassatoquoik; and the icy blue Katahdin Lake, just four miles from Mount Katahdin, the centerpiece of Baxter State Park. The conservation department was outbid, however, and the pristine land near the park faced an uncertain future of likely habitat loss, clear-cutting, and housing. Yet Quimby's purchase next door—the only land that was guaranteed to remain wild—was singled out for criticism.

When Quimby, as expected, declared her land a wilderness sanctuary, open for hiking and human-powered recreation, but closed to hunting, campfires, tree-cutting, snowmobiles, and ATVs, the complaints escalated to something approaching hysteria. Quimby's signs posting the restrictions were torn down. She received threatening and insulting e-mails, letters, and phone messages. Bumper stickers soon appeared: "Ban Hunting: No! Ban Fishing: No! Ban Public Use: No! Ban Roxanne: Preserve the Maine Way of Life." Overwrought news stories detailed the plight of various tenants Quimby had inherited from the previous landowner; the tenants' principal complaint seemed to be that Quimby was not going to let them live on her private land—hunting, logging, and snowmobiling as they pleased—while paying 1970s-era rents. About six months after buying the property, she raised rents to match what other landlords in the area charged. A cabin by the Penobscot River, for instance, was raised from $500 to $1,500—a year. But the sense of entitlement some Mainers had for their "right" to use Quimby's property was powerful. "The people here are very angry," one tenant said. "They feel betrayed." In the state capital, a government task force was formed to look at the issue. Two state senators, a Democrat and a Republican, held a fact-finding meeting in the little village of Shin Pond near Quimby's new land, where local businessmen and residents gathered to complain about her presumed conspiracy "to depopulate the area" and return the woods to nature. Several people urged the senators to persuade state officials to use the power of eminent domain to seize the land from

Quimby. "You can cut her off at the knees," one man exclaimed. The two senators—Paul Davis and Steve Stanley—were uncomfortable at the suggestion, given Mainers' usual emphasis on the rights of private property owners not named Roxanne Quimby. But neither of them would rule out the idea.

The concerns went beyond the cultural bias against a former hippie, now a millionaire, who did not want hunters to kill bears and moose on her property. The underlying fear was for a local economy in jeopardy: the hunting camps and snowmobile trails brought tourist dollars to the rural area, supporting a small market and a hardware store—sources of income that might dry up because Quimby desired to preserve rather than exploit the woods. And although the Shin Pond economy was tiny, if Quimby continued to purchase forestland such "collateral damage" would mount. This was a legitimate fear—but no different from the reality every tenant faces everywhere when a home or a business is put up on land someone else owns, and without any sort of long-term lease or binding promise. Landlords sell. New owners set their own rules. The anti-Quimby forces felt they were defending a way of life, but from Quimby's point of view, she wasn't doing anything unreasonable by determining the fate of her own land.

Quimby did not expect the vilification she encountered—"Doesn't everybody love a park?" she had reasoned—but she was not passive when attacked. Her instinct was to strike back. She pointed out the hypocrisy of her critics who slammed a national park as big government run amok even as they sought to deny her the rights every other property owner in Maine enjoyed. She slammed the governor for failing to back a national park even when Restore's polling showed that a majority of voters statewide supported it (if not so much among residents in the Maine Woods vicinity). Finally, she openly contemplated running for governor with the Maine Green Independent Party, which invited her to be the keynote speaker at its next convention. She later summed up her initial responses to attacks on her land use policies this way: "This land is a treasure made by something larger than ourselves, but there are no guarantees that it will stay this way. I don't think it should be left to chance. And the millions I've put into it gives me the right to determine the fate of the land I paid for. Really, I feel I'm doing what the state should have been doing for years."

So Quimby continued purchasing land in the woods, ignoring the protests. She made it clear that she truly was leaving nothing to chance when she bought two parcels of land near the state park that gave her ownership of the only road for thirty miles traveling east and west. By closing the road to vehicles, she effectively stopped logging in the area. Land acquisition had made her a sort of general, cutting off the enemy's supply lines. It was not a particularly benevolent image, but it demanded respect from adversaries, because the roads could be used as a bargaining chip.

Horse trading—that was something everyone in the forest could understand: If timber companies wanted to use her road, they'd have to pay, in money or, better yet, with old-growth forest in a straight-up exchange. In this way, trading land and access, she acquired more than 65,000 wild, pristine acres in the immediate area of Baxter State Park over a period of three years, and 30,000 acres elsewhere in the woods, spending more than $45 million in the process. Almost all her purchases consisted of lands high on the list that state conservationists and wildlife organizations felt were vital to preserve but had never been able to afford, including 20,000 acres surrounding the Wassataquoik Stream—one of the state's outstanding wild rivers—and the nineteenth-century trails Theodore Roosevelt once used to climb Mount Katahdin. And after each purchase, Quimby systematically closed the land to hunting, snowmobiling, and the particularly destructive all-terrain vehicles, which cut deep ruts in the soft damp soil of spring and spoiled whole habitats very quickly. But hikers, birders, and campers were welcome. She barred use of logging roads that had been cut through her property, and began the long, slow process of restoring the natural contours of the land. Her staff ecologist, Bart

DeWolf, began tramping through the lands, inventorying the plants and wildlife, and he soon discovered rare wild orchids and other vanishing species of plants, which Quimby wanted to protect.

Anger continued to well up with each new purchase, and Quimby could not help noticing a disturbing pattern: Conservationists in other parts of the country and the state would applaud her efforts, but the local people, those who knew and loved the Maine Woods best—people she had lived with for nearly two decades—made it clear that they hated what she was doing. They seemed to hate her, too, or at least to hate what they felt she stood for. Quimby and Restore, it was said, were on the open-season hunting list along with coyotes and other pests. She realized she could buy all the land she could afford, yet she might never achieve her goal of a Maine Woods National Park unless she could persuade the locals that she shared common cause with them. And she believed that beneath the rhetoric, ill will, and seemingly irreconcilable differences, there was more in common than not between her and her critics. She decided to change her approach to see if she was right.

She attempted to broker a deal between factions—the loggers, the state, and the conservationists—by offering the first property she had bought near Baxter State Park, the purchase near Shin Pond that started the uproar against her. She hoped to swap it for one of the properties developers had snatched at the same time, where there were still trees as old as 300 years, very rare in the heavily logged forest. Her first attempted land swap failed as negotiations bogged down in the politics of land use and state budgets. Next she proposed another trade with the same developer, who had threatened to build a new road and bridge over virgin river territory because he could not use logging roads on Quimby's land. This time she left the state out of the mix and dealt directly with the developer. She offered to give him that first big parcel she had bought, where the tenants and snowmobilers were still up in arms—and she would get in return a more pristine wooded property adjacent to the park, where the Wassataquoik

Stream flowed free. He would have his road access back; there would be no bridge and no new access road built; and the tenants, the snowmobilers, and the sporting camp that operated on Quimby's original land purchase would be able to resume living, playing, and working as they had always done. The developer agreed, the deal was sealed, and—this was a new experience—Quimby's critics began praising her for compromising. No faction had gotten everything it wanted in the deal. The developer wouldn't be building on prime forestland as he had intended. Quimby had to give up some land she was attached to. Locals would still have to observe her restrictions—no hunting, no snowmobiling—on the land she had just acquired. But everyone came out with something.

Quimby emerged with a new business model for conserving the Maine Woods: compromise. She stopped talking about running for governor—that was just another CEO job, she decided, and she had had enough of that. She distanced herself from Restore, leaving the board and the park campaign, though she made it clear that she remained inspired by the group's mission and wished it success. The problem was, Quimby said, that Restore was perceived as an outside, interfering Massachusetts-based group of environmentalists having little in common with ordinary Mainers, who liked to hunt and fish in the woods. Quimby thought this was unfair, because the group's operation in Maine is run by a third-generation Mainer with roots in the logging industry, but that's the perception nonetheless, and it was rubbing off on her. So for purely political reasons, she left the organization, and then set about introducing herself as a real person, rather than the cartoon that was the object of the bumper stickers and hate mail. She picked up the phone in October 2006 and called her toughest foes: the town manager of Millinocket in the heart of the wood basket, the heads of the Sportsmen's Alliance and the Maine Snowmobile Association, a state senator who had been her harshest critic. Can we meet, she asked them, face to face? And once they got over the shock that she would call them at all, they agreed to meet.

At the gathering, as her curious guests looked on, Quimby began by saying she knew they had differences: She'd never persuade them to adopt her vegetarian lifestyle, and they'd never convince her that it was OK to shoot animals. Fine. Now let's talk about what we *do* agree on. And she started a list.

We are all staunch defenders of private property rights, yes? We may not agree with what the other guy does with his property, but we have to respect the owner's rights. This was America, after all. Everyone agreed.

And we all love the woods? Vigorous nods. And those woods are changing too fast, and are being diminished in many ways—can we agree on that? Yes, they could. Roxanne Quimby beamed. She presented everyone with a pound of fudge. The people at the meeting were suspicious, uncertain, but Quimby had lived in a cabin in Guilford and had endured the harsh winters that Mainers know so well, the penetrating cold that sears any exposed flesh. She talked their language. She was no longer a rich witch with the millions from a cosmetics empire. She was the woman who had to borrow the town shovel in Guilford when it snowed. And by the end of the meeting, without going into a great deal of specifics, they had agreed that there were probably some compromises they could strike in which all would get something they absolutely had to have, and they all could give up some things they might want but didn't absolutely need. Important conservation could happen in the Maine Woods, responsible hunting and recreation could continue, what was left of the forest industry could still operate in the younger, well-used parts of the forest. Everyone could win. And as a show of goodwill, Quimby agreed to keep open a prized snowmobile trail on her newest property acquisition for a full year rather than close it immediately as she had intended. And after that, they would see.

The meetings soon became a monthly fixture, each of the members of this informal working group taking a turn as host. Quimby remained straightforward about her goals. She wanted to continue to establish wilderness sanctuaries that, for the most part, did not allow hunting, snowmobiling, or ATVs, and that she hoped someday would form the core of a single new park or protected wilderness. But she would be flexible, she promised, and make some areas open to traditional uses. The rest of the group members appreciated the clarity this brought to the process: Everyone knew where Roxanne stood. Pretty soon, Quimby had become the lesser evil—a person they could deal with, rather than the state bureaucracy or a nonprofit organization with a board and donors to answer to. Quimby could make a decision on the spot; she answered only to her own conscience. "You know, Roxanne," one of her former critics said after a year of meetings, "you're a real redneck." It was a compliment—a statement that she was one of them. A short time later, a new agreement was publicly announced in which Quimby was able to buy another 9,000 acres of prime forest land near her existing holdings in the Wassataquoik Stream region—land the state wanted but couldn't afford on its own. The purchase nearly completed her plan for amassing a huge swath of contiguous territory east of Baxter Park for permanent preservation. At the same time, she agreed to grant easements for recreation and logging on a 7,000-acre area near Millinocket, land that had low conservation value but important roads for logging and two favorite snowmobile trails. As a result, Quimby's most vocal detractors appeared with the governor at a press conference, praising her as a good neighbor and steward of the land.

"Roxanne deserves a great deal of credit for bringing together groups with a very different vision from her own," Gene Conlogue—the Millinocket town manager and an outspoken opponent of the park—said at the press conference. "And getting all of us to roll up our sleeves and develop a solution that works." Newspaper columnists wrote in amazement about the "thaw" in the Maine Woods that just a year earlier would have seemed inconceivable.

Quimby had also accomplished something much deeper than just placating critics and securing more conservation lands. Her lands were becoming a de facto park. They lacked the infrastructure of a park, but the most onerous part—the restrictions on hunting, logging, and vehicles—was there. And the world hadn't ended. The restrictions existed in a framework people accepted, because everyone got something. She had succeeded in making the idea of the park real. Restore had never been able to do this as an advocacy group with no land of its own. A resort operator on Roxanne's former land at Shin Pond, previously angry but subsequently satisfied, talked of making a "180-degree turn" in his thinking about Quimby and her mission. Then, unwittingly, he quoted almost verbatim Quimby's thoughts when she first convened her monthly working group: "We love our land, and maybe, in the long run, we all want the same thing."

Far from feeling jilted because of Quimby's departure from Restore and its campaign for a national park in Maine, Jym St. Pierre was ecstatic. In quitting the group so many Mainers loved to hate, Quimby had done more than anyone else to advance its cause, and for one simple reason. On her own, she was able to make the park idea tangible. The land was there. The critics had embraced the idea. It was a start.

Around the time of the "great thaw" between Roxanne Quimby and her detractors, the Land Use Regulation Commission held a final round of hearings around the state on Plum Creek's proposal for a housing and resort development at Moosehead Lake. The purpose of the hearings was to solicit public comments on the proposal that would determine the future course of development and preservation in the Maine Woods. One of the best-attended and most impassioned hearings was at Greenville High School in January 2008. Men and women of the Moosehead Lake region gathered to discuss the future of a landscape which they had always thought was eternal, but which now would change, one way or another.

"Saving Moosehead Lake is a cause for a lot of people," testified Pinkie Bartley, a resident of Greenville for fifty-two years and a proponent of approving the Plum Creek proposal without delay. "They'll come up here for today and tell you how much they love this place and how it shouldn't be changed at all—and then they'll leave. Meanwhile, the rest of us are here to stay. We're not fighting for a cause. We're fighting for a say in our future."

To some, that future looked ominous without Plum Creek in the picture. Stacy Fitts, a local legislator who sat on the Select Committee on Maine's Future Prosperity, predicted the demise of the region's economy if the plan was rejected. "The Moosehead region is crying for a change. This is the only place I have ever seen where a McDonald's sits boarded up. It's amazing to look at when you pull into this town." Others spoke of the recreational opportunities that would be preserved with Plum Creek's development, but that might vanish if the project was rejected. "I've hunted and hiked and fished and kayaked and cross-country skied and snowshoed. . . . To me, as a non-landowner, to have that opportunity is just fantastic," the attorney Jeff Cummings argued. "We know from the Roxanne Quimby issues what can happen with certain landowners. And if this plan isn't approved and if this land is subsequently sold to another landowner, we know that a lock can be put on the lands, literally. They can be shut down."

A highlight in the case in favor of Plum Creek came when a local historian, Elaine Bartley, displayed photos to prove that in the past, development outstripped even Plum Creek's modern proposal, though nature has largely reclaimed the Moosehead area since then. Lilly Bay had a resort hotel as recently as the 1930s, Bartley explained, built on the foundation of a lumber workers' lodge—just as big as the one Plum Creek now proposes. Rail service was extensive a century ago, far better than it is today, with trains from New York direct to Kineo Station near the now protected Mount Kineo, the jut-jawed promontory overlooking Moosehead Lake. In 1892, the Coburn Steamboat Company built a wharf next to the train station and ran a fleet of lake steamers serving lodges, hotels, and camps all around the lake.

Yacht races and regattas were daily events in the tourist season, and the Kineo Hotel held 400 guests and fielded its own baseball team. For a moment, as Bartley showed off the old photographs, the ghosts of Greenville's and Moosehead's past glory filled the meeting room, echoes of a more genteel age—when train and carriage were the principal methods of transport, global warming was unheard of, and the forest seemed as limitless as the technological progress promised by a new century, the twentieth. "An opponent . . . said that approving the plan would be like breaking up a masterpiece painting," she said. "I think just the contrary. Approving this plan will give us the opportunity to put back our picture of our history."

Noel Wohlforth, a member of the school board, was among those who thought the masterpiece was under siege and that Plum Creek needed a new and better plan. "What in the world would a plan like this do to this beautiful town of Greenville? It scares me. The term in here that I found the most startling is 'Impact Zone.' We are an impact zone. When I was in the army that meant a bomb was going to drop. I kind of think in a way this is the bomb."

David Foley, a former town planner, decried Plum Creek's "dark hints" that a denial of its proposal would spell doom for the forest and the region. "You could almost admire their audacity if the consequences weren't so devastating. . . . Plum Creek thinks we're a bunch of rubes who just fell off the turnip truck. When you make sense, you don't need battalions of attorneys, spin doctors, and fixers. When you're a straight shooter, you don't need bamboozlement and not-so-subtle threats."

Stephen and Meridith Perkins spoke almost poetically of a small camp they have on Mud Cove on the east side of Lilly Bay, of the lynx they have seen near their cabin, of the sound of the loons on the lake at night, and the answering call of the coyotes, a conversation that has been going on for thousands of years, and that will probably be lost if Plum Creek's plan is approved. "Are you willing to trade the call of the loon for the ring of the cash register?" Stephen Perkins asked.

Perhaps the most eloquent comments at the hearing from either side came from Jayne Lello, a teacher from the town of Sebec, who had prepared for the hearing by reading from Thoreau's *The Maine Woods* and C. A. Stephens's 1888 *Knock About Club in the Woods*. The testimony of citizens at these public hearings is seldom reported in the news media and rarely read—the transcripts sit and molder as the wheel of government turns to the next crisis—but Lello's simple, loving words of caution and her plea to reexamine the proposal, not to kill it but to strike the best possible balance between nature and commerce, are worth noting:

> *My thoughts on this whole issue are relatively simple, boiled down like maple syrup to two words and one question. How much?*
>
> *We know that things are going to change. And LURC's task as our state crossing guard is to answer this question. We know Plum Creek must and will make a profit. How much?*
>
> *We know some wild and special places will be forever lost when this deal is made. How much?*
>
> *And there will be some economic gains and losses for this region. How much? I can only imagine the smile on the face of Plum Creek's land prospector when he or she first flew over the incredible landscape realizing it was all for sale. How much?*
>
> *At zoned timberland prices, we know it was pretty much a steal. Now, not surprisingly, they want to rezone those acres from timberland to wonderland prices. How much?*
>
> *Maine people wonder what the fallout and what the advantages will be. Everything will change, but how much?*
>
> *My husband has always said that Maine is like Vermont's poor cousin. Vermont, I think, has always realized it's worth more than Maine. It has good self-esteem and laws that protect its unique land*

and lifestyle. Too often I fear that Maine sells itself short. We're a
poor state, but we are rich in resources—water, forests, mountains,
and coastline. Maine people understand and appreciate the incred-
ible resources outside our backdoors just beyond our woodpiles; but
we can rarely afford the real estate prices. We are very proud and
feel a responsibility to what happens here because we know these
ponds, lakes, streams, woods, back roads, and mountaintops. We
don't own them personally; but like a common-law marriage or
an old familiar habit, we are tied together, Mainers and the land.
The strong character of the people here has been chiseled from
the forceful nature of the place. This Moosehead region with its
breathtaking views, slippery sparkling fish, running waters, swift
deer, lumbering moose, craggy mountaintops, and ever-resilient
forests always takes our breath away and never fails to rekindle our
sense of wonder and our exuberance for life. That's why we're here
and why we stay. Historically the native people who named these
landmarks, Henry David Thoreau, C. A. Stephens, Holman Day,
and on to eloquent writers in this time, are amazed and inspired. It
doesn't discriminate between natives and visitors. It enriches us all.

"How much?" is the main question.[1]

When the comments from the public hearings, along with written testimony mailed and e-mailed, were tallied, it became clear that the terms of the debate had shifted, that Lello's questions were shared by many, and that the compromising approach of Roxanne Quimby was working against Plum Creek and the specter of mammoth develop-ment in the Maine Woods. The comments ran twenty to one in call-ing for the commission to reject the proposal and to send Plum Creek back to the drawing board. It seemed likely that many more months, possibly years, would pass before the issues were resolved and any sort of project could proceed, and that even after LURC made its decision, there would almost certainly be time-consuming lawsuits and endangered species petitions and continued debate. Some sort of development would eventually take place, but it would no longer preclude significant land preservation in the woods.

Quimby, meanwhile, continued to buy up land, and her ability to do so increased significantly in November 2007, when the Clorox Company, best known for making bleach, bought Burt's Bees for $913 million, with a promise to continue its green policies (and to incorporate at least some of them in other divisions of the company). Quimby's remaining 20 percent share was worth $182 million—more than she had made from selling the entire company just four years earlier. Quimby acknowledged that the buyer had not previously been known for environmentalism—and that its signature product, bleach, was bad for the environment—but she felt the sale would allow her to accelerate her conservation efforts in the Maine Woods, and to expand her efforts to other areas as well. If selling the company created a karmic debt, she would later say, her use of the money for conservation settled the account.

In addition to continued purchases in the forest, Quimby began buying—acre by acre—the in-holdings in Maine's Acadia National Park and other national parks around the country. In-holdings are small pieces of privately owned land inside national parks that were retained by the original owners when the parks were formed. Now and then these holdouts come up for sale, and sometimes developers snap them up, then threaten to erect an inappropriate mansion or some other structure that could spoil the ambience or views within the park—a strategy designed to coax an exorbitant purchase offer from the U.S. National Park Service. Quimby decided to try to swoop in and buy in-holdings instead, sometimes acting through intermediaries with sellers who might not approve of her conservationist ways. In 2016, on the hundredth anniversary of the national park service, she says she will present a portfolio of in-holdings to the superintendent of Acadia National Park, and another to the park service nationwide. She plans to donate the land with no strings attached.

And on that anniversary, she hopes to propose something else with the land she has now and plans to obtain in the future in the Maine Woods—the same thing Franklin D. Roosevelt and John D. Rockefeller Jr. did in the face of opposition to the creation of a Grand Teton national park. She will suggest a national monument. It is the perfect solution: less than a park, but still protected, and it requires only the order of a president, not an act of Congress.

"We're not quite ready for a park," she says. "A monument or a national wilderness area is less threatening. It's doable. And it will protect nature. And once it's there, and people find they can live with it, enjoy it, and prosper because of it, then turning it into a park becomes no big deal. That's what I'm gunning for. And that's what will save the Maine Woods."

Part Four

lone wolves

It Doesn't Take a Village (or a Billionaire,
Although a Billionaire Doesn't Hurt)

*Consider this: all the ants on the planet, taken together,
have a biomass greater than that of humans.
Ants have been incredibly industrious for millions of years.
Yet their productiveness nourishes plants, animals, and soil. Human
industry has been in full swing for little over a century, yet it has brought
about a decline in almost every ecosystem on the planet.
Nature doesn't have a design problem. People do.*

—WILLIAM MCDONOUGH AND MICHAEL BRAUNGART,
CRADLE TO CRADLE: REMAKING THE WAY WE MAKE THINGS

*I'd put my money on the sun and solar energy. What a source of power!
I hope we don't have to wait until oil and coal run out before we tackle that.*

—THOMAS ALVA EDISON, 1931, SHORTLY BEFORE HIS DEATH,
TO HENRY FORD AND HARVEY FIRESTONE

andy frank and the power of the plug

Professor Andy Frank has a question for you:

What if there were a full-size sedan or SUV that performed like a sports car, cut gasoline consumption and greenhouse gas emissions by 90 percent, burned no fuel at all for the first sixty, eighty, or hundred miles—and cost about the same as a regular car?

"Everyone would want that car," Frank says with a laugh. "People would go crazy for it. It would save them so much money. It would save us from oil dependence. It would go such a long way toward saving *us*, while preserving our way of life."

Sounds like a very compelling vision for a car of the future, perhaps something to shoot for in the next ten or twenty years—except, it is not a vision. The car is sitting in Frank's garage right now, or it is parked at his university, or tooling around the campus. One such car was even hidden somewhere in the bowels of General Motors years ago, after the company hired Frank to build a prototype, then made the stunning results disappear. Indeed, Andy Frank, a legend among designers of alternative-energy vehicles, has been building such cars for twenty years, starting when he was fifty-eight. He's been battling a recalcitrant industry and confused policy makers to bring the cars into the mainstream ever since.

Frank invented and patented the plug-in hybrid—a vehicle with both electric and gas power sources that can also be plugged into a standard electrical outlet to get those sixty or more gas-free miles. His concept, quite simply, can help save the country and the world from oil dependence, skyrocketing energy costs, and environmental ruin. "It will happen," he says with surprising serenity, given his long years of frustration. "It will happen because the price at the gas pump will finally make it happen."

At his warehouse and lab at the University of California–Davis, the professor of engineering and his students take inefficient Detroit iron and turn it into green miracles—vehicles powered by big electric motors and small, gas-sipping engines the size of a motorcycle's, which nevertheless outperform the originals. His years as a teenage hot-rodder show in his insistence on not sacrificing performance for efficiency: His modified sport utility vehicle can get zero to sixty in ten seconds, three seconds faster than when it left the factory floor. Yet the cost of operating his customized cars would be about the same as a regular vehicle's gasoline costs—if it were still 1978, that is, and gas still cost seventy cents a gallon. That's what comparatively cheap electric power and using hardly any gas can accomplish: Frank's cars can cover so much more distance than a standard car on the same amount of gasoline that it's as if his fuel cost seventy cents a gallon.

Frank's plug-in hybrids are models of efficiency, from their dearth of moving parts to their onboard computers to the regenerative brakes that make electricity as the car stops rather than wasting energy through friction and heat. His revolutionary design for a continuously variable transmission that merges the electric and internal combustion power into a seamless drive train has only twelve

moving parts; the standard Detroit automatic transmission has 700.
The plug and a beefy battery pack tucked under the floorboards separate Frank's cars from standard hybrids available to consumers as of
2009, such as the wildly popular Toyota Prius. The Prius improved
efficiency by using both gas and electric motors; it got forty-seven
miles per gallon, making it the most fuel-efficient mass-produced car
on the market in 2008. But it had no plug (yet), so its range as an
electric-only car was very limited. The sixty-mile to hundred-mile
all-electric range Frank builds into his plug-in hybrids is crucial to
their green credentials and allows them to drive rings around the
Prius: Because 80 percent of Americans drive fifty miles a day or
less, Frank's technology allows most drivers to travel to work or to
the store or to drop the kids off at school entirely without burning
fossil fuels and without visiting the gas station except to pump up the
tires. Using a modest-size solar collector, as Frank does, to provide
the plug-in power makes the system completely carbon-neutral, and
eliminates the cost of charging the batteries. One of the main sources
of climate-changing greenhouse gases—car travel and its dependence
on oil—would be nearly eliminated if America and the world drove
Frank's plug-in hybrids. And the car's fuel costs, once the solar panels
were paid for, would be zero.

If the automotive industry adopted ideas on their merit alone,
Americans would already be driving plug-in hybrids en masse, but
other factors—corporate fear of change, the auto industry's unwillingness to appear to embrace emissions regulations, and the oil industry's unsurprising opposition to weaning Americans from oil—led
Detroit in the 1990s to bet on inefficient, dirty, but popular sport
utility vehicles rather than Frank's unproved, clean hybrids. That
doesn't mean the industry doesn't take Frank seriously: You can't
ignore an inventor who uses a bunch of college kids and some off-the-
shelf parts to turn a Ford Explorer into the efficiency equivalent of a
100-mile-a-gallon motor scooter. For decades, carmakers in America
and Japan, as well as the federal government, have hired Frank, con-

sulted with him, given him grants, capitalized on his ideas, presented him with awards, studied his designs, and had him build prototypes that worked fabulously, accomplishing everything he promised—and this has been his downfall. His cars are too good.

Oil companies, he says, don't want a world where drivers visit the gas station once every two or three months—or less. That's why Frank believes the oil industry is so eagerly backing the hydrogen fuel cell car, endowing university programs to study it, and persuading government to support it—because it is a technology that is decades in the future. And, he says, if it ever matures, the oil companies would still control the fuel supply. The allure of fuel cells is their theoretical cleanness: They chemically combine hydrogen and oxygen to produce electricity, with water vapor as the only tailpipe emission. There are several catches, though. At present, the only practical method of extracting large amounts of hydrogen for use in cars is from dirty fossil fuels—hydrogen's clean image is just that, an image. Further, making hydrogen requires large amounts of electricity—as much as four times the amount of power that the hydrogen produces in a fuel cell. As Frank puts it, he can power four of his plug-in hybrid electric cars with the same amount of electricity one hydrogen car sucks up each time it is refueling. And refueling is a problem, too—even if fuel cell cars were mass-produced, there is no national infrastructure for shipping, storing, and pumping hydrogen into people's cars. Finally, hydrogen fuel cells are expensive and use scarce materials, such as platinum. Yet American car companies are only too happy to go along with this idea. Why should carmakers retool their assembly lines for a switch to plug-in technology that exists today, Frank says, when they can appear to be good environmental citizens by researching the distant possibility of hydrogen and providing some expensive prototypes for movie stars and corporate chieftains to drive?

"They like hydrogen, because it's thirty or forty years away," Frank says with a dry laugh as he scrambles around the basement garage of the Petersen Auto Museum in Los Angeles, where he and his university crew are showing off their latest prototype, a hybridized Chevy Equinox code-named Trinity that can handle a sixty-mile commute without burning any gasoline. "You know what we say in this business: Hydrogen is the fuel of the future. And it always will be."

Frank is probably the most patient inventor this side of Thomas Edison, who famously tried 6,000 prototype lightbulbs before finally coming up with a practical model. But now, at age seventy-five, although he looks twenty years younger, Frank says he's done with waiting. Between pressure on government and industry to take firm steps to alleviate climate change, and the price of gasoline hitting all-time records in 2008 before a recession sent prices tumbling, Frank says his time has come. More than ever, he is in demand as a speaker and consultant, and it appears that some car companies are moving forward with plug-ins. Several models are scheduled to be on the showroom floor by 2010, although they fall far short of the performance Frank has proved is attainable. China may be the furthest along, setting the pace in a new industry that Detroit could have commanded fifteen years ago.

"They're taking baby steps, but at least they are stepping in the right direction nonetheless," Frank says with a small smile that quickly fades. "But I have one worry: that it's coming too late."[1]

Andy Frank has been saying for years that oil is not a need but an addiction—with all the deadly, irrational, and self-destructive behavior that the word "addiction" implies. People used to react as if that were crazy talk, but it is an idea that has gained currency. Even America's oilman president, George Bush, started talking about the nation's addiction to oil toward the end of his time in office, though, perversely, his prescriptions for dealing with that dependency always revolved around making the addiction and its consequences worse, not better. Bush's economic stimulus package, adopted by Congress in 2003, summed up the nation's energy policies at the start of the twenty-first century: It left Americans with a maximum tax credit of $4,000 for buying a zero-emissions electric car, versus a $100,000 tax deduction for buying a vehicle over 6,000 pounds, which included the ten-miles-per-gallon GM Hummer H2. By 2008, the deduction for the Hummer had dropped to $50,000; but with a base price of $52,000, that meant taxpayers could foot almost the entire bill for the purchase of one of the most fuel-inefficient, greenhouse-gas-spewing cars in the world—the ultimate subsidy for Americans' oil addiction.

The 2008 subsidy for buying an electric car that burns no gas at all, that weans its owner from addiction, and that emits no greenhouse gases: zero. (Hybrids with no plugs were granted a tax credit of up to $3,400; but in practice, the credits for most models were much lower, and credits were phased out for the most popular and most fuel-efficient hybrids.[2])

It is this sort of incentive—resulting from a combination of federal policy, industry lobbying, partisan politics, corporate malfeasance, and consumers' hostility to change—that has made oil, and specifically gasoline, the most addictive substance on the planet, and has kept it at the center of our economy and our lives long after it could—and arguably should—have been relegated to a minor role as an energy source.

Addiction's cruel power lies in creating the illusion that the addict cannot survive without the desired substance, be it heroin, nicotine, or gasoline. Given the omnipresence of oil-powered systems in American society today, with the most toxic mode of transportation ever conceived embedded in every aspect of our daily lives—mowing the lawn, doing the shopping, going to work, doing our work—it is almost impossible to imagine an alternative. The automobile and oil industries, often aided by government, reinforce this perception by continually making the case that there is no viable substitute if we are to sustain the American way of life. This apparent truism has been used to oppose every regulatory effort in the last four decades to replace or improve the automobile in order to reduce smog and (more recently) to reduce global warming, to improve efficiency, or to make cars safer, just as manufacturers opposed regulations requiring seat belts and air bags.

The argument of necessity is so convincing—indeed, so seemingly evident—that it is easy to forget that neither gasoline nor the internal combustion automobile was an American invention, nor was either responsible for most of civilization's great achievements. Without them, humanity somehow managed to invent democracy, geometry, physics, atomic theory, subways, representative government, Hammurabi's code, the Ten Commandments, Beethoven's Fifth, the Magna Carta, the Declaration of Independence, the United States Constitution, the theory of evolution, the Stradivarius, the printing press, ocean liners, locomotives, calculus, clocks, vaccinations, gunpowder, rockets, eyeglasses, pasteurized milk, beer, contact lenses, telescopes, microscopes, generators, lightbulbs, batteries, radio, telephones, and chocolate, and also managed to determine that the world was round, to circumnavigate the globe, to settle the western United States, to lay the transatlantic cable, to reach the north and south poles, to scale the Matterhorn and Mount Kilimanjaro, to abolish slavery in Europe and America, to electrify the cities of the world, to record music and movies, to discover the principles of genetics, and to build an American electric rail system of trains and trolleys with nationwide reach that, at the beginning of the twentieth century, carried 12 billion passengers a year.

Gasoline, then an insignificant petroleum by-product—often simply burned off as waste during the production of kerosene for lamps—was sold in small bottles to treat lice infestations in children's hair.

At the end of the nineteenth century, among the earliest and most popular cars were battery-electric vehicles, and by 1900 they held both speed and range records over rival external combustion steam-driven cars, and over their new internal combustion, gasoline-burning competitors. Early electric cars were the first to reach a speed of a mile a minute, and the first to achieve speeds higher than 120 miles an hour. The Electric Storage Battery Company (forerunner of the battery giant Exide) operated a fleet of electric cabs in New York City, with a system for quickly swapping exhausted batteries for charged ones so cabs could be quickly returned to service. The first car race on a track in America, held in 1896 at Narraganset Park in Rhode Island, pitted five gasoline-powered racers against one electric, the Riker, in a one-mile sprint. The electric car won. Even Henry Ford, who collaborated with Thomas Edison as the famed inventor developed a state-of-the-art battery for powering autos, publicly expressed the belief that electric cars needed to play a big part in motoring into the twentieth century. As he launched the most popular and revolutionary car of the age—the gasoline-powered Model T, which debuted in 1908—Ford chose electric cars for his wife and son.

Electrics were clean, quiet, and mechanically simple, and they were extremely efficient—about 90 percent of the energy put into their batteries ends up being used to propel the car, with only 10 percent wasted. Gasoline cars were loud, noisy, mechanically complex, emitted clouds of toxic exhaust, and were—and are to this day—incredibly inefficient. In early cars, as little as 13 percent of the energy produced by burning gasoline was actually used to propel the car—the rest was lost through waste heat, friction, throttling, idling, and powering ancillary systems such as the water pump. Modern cars do better—they manage to use about 20 percent of the energy from gasoline while wasting 80 percent. (This means that out of $4.50 a gallon—the cost of gas in the summer of 2008—as much as $3.60 was being blown into the atmosphere, while only ninety cents went toward moving the car.) At the turn of the twentieth century, electrics were as common as gasoline-powered cars, accounting for nearly one-third of all car sales. But by 1905, the tide had turned: Electrics accounted for only 7 percent of new car sales, whereas 86 percent of cars bought in America that year were gasoline-powered. The dirty, inefficient technology beat out the clean, efficient technology, with lasting consequences—for our environment, our economy, and, once our oil addiction impelled us toward foreign suppliers, our national security. But this was a choice, even though it has come to be viewed as inevitable.

The common explanation for the triumph of internal combustion vehicles over electric vehicles is that market forces went to work, consumers considered the options, and the better car won. According to this view, early internal combustion cars and gasoline, even with their admitted drawbacks—noise, pollution, and unreliability—made for faster, more capable cars with longer ranges. Batteries of the era are said to have been too heavy and to have performed poorly, making electric cars painfully slow and limited in range, requiring a recharge after only a few miles, whereas gasoline cars kept chugging on down the road. This conventional wisdom about electric cars also appears in contemporary objections by the oil and automotive industries: poor batteries; short range; market forces know best. But the conventional wisdom on this historic market decision is wrong, and the conditions that made gasoline cars the logical choice in 1910 no longer exist today.

Internal combustion won less on its innate merits and more on cost and infrastructure: Electricity was expensive in 1900, and gas was cheap. In 1900, electricity cost twenty to forty cents a kilowatt-hour—so it was far more expensive in 1900 than it was in 2000 (even without considering inflation). By contrast, gasoline was a nickel a gallon at the beginning of the twentieth century. (The price differential is even more stark when we adjust for inflation. In constant 2007 dollars, electricity a century ago cost the equivalent of up to ten dollars a kilowatt-hour, compared with contemporary rates of ten cents for the same amount; gasoline prices a century ago were the equivalent of $1.27 a gallon in 2007 dollars.) With the lead-acid batteries in use at the time, it would cost the 2007 equivalent of twenty-five dollars to travel fifty miles in a 1900 electric car, whereas a gasoline powered vehicle could do it for less than four dollars.

Availability also favored internal combustion. Gasoline was quickly and easily stocked by stores all over the country, and a 1913 gas station building boom met (and increased) demand. In comparison, electrification was slow. Most Americans couldn't afford to have their homes wired early in the century; electricity in the home was not commonplace until after World War I, when rates declined and availability increased, and even then it was mostly limited to cities. Rural electrification did not begin in earnest until the mid-1930s. But claims that the technology of electric cars was inferior appear to have been more propaganda than reality. The top speed of one model popular early in the century, the Baker Electric, was thirty-five miles an hour, about the same as the most popular car of the era, the Ford Model T. The Baker could travel fifty miles on a single charge, and 100 miles with Edison's new batteries. In theory, gasoline-powered cars could go much farther on a tank of gas (the 1908 Model T got better miles per gallon than the 2008 Ford Explorer), but in practice, early gasoline cars tended to overheat or break down and on average needed to pull over every twenty miles or so to be worked on.[3] If anything, the effective range of early electric cars, as well as their reliability, was superior.

It is true that early lead-acid batteries—the same technology used today to start gasoline engines, though current versions are significantly improved—were not well-suited to the rigors of providing the main power for vehicles. They were heavy, bulky, and filled with corrosive chemicals. Edison and his battery team conducted thousands of tests on numerous different battery chemistries—reminiscent of his long trial-and-error search for the perfect incandescent lightbulb filament—before perfecting, in 1910, a nickel-iron alkaline battery that he believed would prove decisive in persuading consumers to choose electric cars. These nickel-iron batteries seemed never to wear out, were much lighter in weight, and could be charged twice as fast as lead-acid batteries, while offering 233 percent more power output. Most important, they offered a seemingly spectacular range when used in a car. In one of Edison's tests, a passenger car traveled 244 miles on a single charge—an accomplishment that would still be considered a breakthrough today. Henry Ford boldly announced to *The New York Times* in January 1914 that he and his friend Edison would soon be producing a new line of affordable cars, costing the same as his Model T, powered by Edison's revolutionary battery. "The car we propose to build will contain a battery equipment weighing 405 pounds and the entire car will weigh but 1,200 pounds. It will run for 100 miles. The cost will be about $600 to the public. How does that compare with the great, heavy, and expensive electric cars?"[4]

Edison and Ford envisioned a network of curbside charging stations, in which a coin-slot mechanism would deliver a metered electric flow—General Electric had actually introduced such a device, calling it an "electricant," or electricity hydrant. Edison had a grand vision of combining the electric vehicle with another new product, a portable home power plant. Driven by either renewable wind power or a small gasoline motor, the personal power plant would charge a pack of his new batteries and thereby power an entire business, a house, or—with several plants hooked in series—a block of houses. Edison argued that these mini power plants, small enough to be kept in a cellar or toolshed, would be cheaper and easier than building large central stations and thousands of miles of electrical grids. The addition of windmills would decrease the need for fossil fuels—and provide free charging for the Edison-Ford car. The vision was brilliant and technically feasible; Edison powered a large house in West Orange, New Jersey, as a prototype, not far from his own home. "The Powers of Darkness have suffered another rout. Thomas A. Edison, their implacable and indefatigable foe, has devised a final scheme for their undoing," the *Times* exulted in 1912.[5]

Edison also attempted to impress the world with his new car battery as well, arranging for a 1,000-mile endurance ride with two electric cars: a Bailey Electric and a Detroit Electric. Each car was equipped with his new batteries and each routinely traveled more than 100 miles on a charge as they drove through four New England states and climbed Mount Washington. *The New York Times* reported in its headline, "Edison Battery Is No Longer a Myth."[6]

Had this vision of the "Edison Suburban Residence for the Twentieth Century" coupled with a practical electric car taken hold 100 years ago, the world would look very different today—and the fast-approaching catastrophe of global warming easily could be a far more distant concern. Edison's and Ford's ideas for the early twentieth century are very similar to the prescriptions scientists and environmentalists now say we desperately need in the twenty-first century. With the addition to Edison's home power station of solar cells—which were mere novelties in Edison's lifetime, though he saw great promise in them—the parallel would be complete.

But it was not to be. The idea of home power plants did not catch on, as the initial investment of $500 for the smallest of the devices and $3,000 for the largest was beyond the means of most households at the time. Ford and Edison's electric car project fell victim to a series of delays and misfortunes: manufacturing and quality-control issues; a devastating fire that swept through Edison's factory and laboratory complex; then the advent of World War I, bringing a navy contract that monopolized Edison's entire battery production capacity. By the time the war was over, the moment had passed.[7] Electric cars powered by old and sometimes shoddily made lead-acid batteries had by then tainted the marketplace, as did the unscrupulous practices of some electric car businesses.[8] By 1920, just as electricity costs began to drop, the major car manufacturers, including Ford, abandoned any remaining plans for electric cars, along with some promising experiments with hybrids.

A similar pattern—based on factors that had nothing to do with technological merit—devastated the extensive electric trolley and interurban light rail systems that commanded billions in ridership in the early 1900s. In 1921, when Americans wanted to go somewhere, they used rail 90 percent of the time, mostly electrical rail systems and streetcars. Only one in ten Americans owned a car in 1921. And the reason was simple: Almost every town in the country with more than a few thousand residents had a trolley service, and together, the nation's 1,200 local and regional electric light rail systems covered some 44,000 miles. Mass transit was convenient, cheap, and plentiful.

That same year, General Motors suffered a loss of $65 million, and the company decided—quite logically—that trolleys and light rail were part of the problem. Converting those clean electric systems to mass transit that used General Motors' fossil-fuel buses became a serious and sensible business priority. A front company called National City Lines—backed by General Motors, Mack Truck, Firestone Tire and Rubber, Standard Oil, and Phillips Petroleum—began buying up the nation's trolley and light rail lines in the 1930s and 1940s, shutting them down and replacing them one by one with gasoline and diesel-powered buses. The collusion was so blatant that the companies were indicted in 1947 and later convicted of conspiring to monopolize trade and commerce in the purchasing of buses, tires, tubes, and petroleum products—a plot, according to the lead federal prosecutor on the case, "to deprive the American public of their splendid electric railway systems."[9] But the $5,000 fine ultimately imposed in no way deterred the immensely wealthy corporate conspirators, and their goal was achieved.[10] Major cities such as Los Angeles saw irreplaceable light rail and trolley lines closed down, the rights of way abandoned and built over, where once they connected all the suburbs, beach towns, and downtowns throughout Southern California. Only a few cities, such as Boston, Philadelphia, San Francisco, and Chicago, held onto their old electric lines. Retired trolley cars were stacked up in junkyards and fully functional electric passenger vehicles were scrapped. Half a century later, Los Angeles would spend billions of dollars trying to re-create a few segments of the old light rail system it had so cavalierly abandoned in the name of modernization.

The failure of the electric car, of Edison's home power plant, and of trolleys and light rail was not inevitable or even sensible. Such technology and infrastructure could have saved the nation and the world from the spiraling oil prices and climate change of the twenty-first century. Instead, the marketplace did what it was supposed to do—maximize profits for investors. This is the real magic of the marketplace, and in many instances, that magic has served the country well. But it is not the role of the market, or the business model of General Motors or Standard Oil, to safeguard the environment voluntarily or to make wise decisions about public transit, particularly when such efforts might diminish profits. In the absence of incentives or mandates from the government to push the private sector toward serving the greater good, it is madness to expect such a balance. This is why the United States trails behind the rest of the industrialized world in combating climate change; why the nation is pursuing corn-based biofuels that industry loves but that will produce even more greenhouse gases than the fossil fuels they are intended to replace; why even China has more rigorous fuel economy standards for its cars; why immensely profitable oil companies get more than $100 billion a year in corporate welfare and subsidies; and why behemoth sport utility vehicles get generous tax breaks while nonpolluting cars get little or nothing.

The most sweeping attempt yet to break with this history by creating a new mandate for clean vehicles occurred in the state with the nation's worst air quality, California. In 1990, government officials decided there was only one way to beat the smog that was causing increased cancer rates and record amounts of childhood asthma, and literally eating away historic buildings: order carmakers to build clean cars. And by clean, California meant cars with *zero* emissions.

California wanted a ZEV—the zero emissions vehicle—for which there was only one possible option at the time: the return of the electric car.

In a grand experiment of market manipulation, the major car companies would have eight years to design and deploy electric cars. At first they'd have to put out only a few hundred. But by 1998, 2 percent of the cars they sold in California had to be ZEVs. In 2001, the figure would become 5 percent; in 2003, 10 percent. At that point, there would be 150,000 clean, cool, low-cost, high-performance electric cars on the road, and the members of the powerful California Air Resources Board figured market forces would do the rest. Who wouldn't want a car that cost only one-fifth as much to operate as a regular gasoline-powered car, and has virtually no maintenance costs?

The mandate came with an incentive to carmakers: Meet the ZEV deadlines, they were told, or they would be barred from selling any cars in California, the biggest car market in the western hemisphere.

The industry criticized the mandate as unrealistic and too costly, and the major carmakers sued to block it. But the protests were not, at least at first, convincing anyone, because the idea for the ZEV program began not with government officials, but with the leading carmaker in the world at the time, General Motors. The company had boasted of developing a revolutionary electric car, a breakthrough in technology and design. At the annual Los Angeles Car Show in 1990, the chairman of GM himself touted the new electric car as the wave of the future. He personally demonstrated a prototype battery-powered car called the Impact, with the sleek look of a sports car and the ability to travel 120 miles on a single charge. "We are going to put the Impact into production. We're dedicated to doing it," an excited Roger Smith, then GM's chairman, announced. The California air board, desperate for a solution to smog, understandably took Smith at his word, and the idea of mandating the ZEV was born. If GM could do it, every major carmaker could, the air board reasoned.

And the carmakers did build the electric cars. They did so reluctantly but well. General Motors' prototype Impact went into production as the EV1. Ford had the little bubble-shaped Think! Toyota converted its smallest SUV into the RAV4-EV, which had a 150-mile range on a single charge, using relatively low-tech batteries—nickel metal hydride models not very different from Edison's old batteries. There were others, too, though the EV1 and the RAV4 were the most popular, beloved by the few who could get them. There were waiting lists with thousands of names.

The new electric car divisions that were established within the major car companies may have done a superb job in creating these vehicles, but this did not seem to please the leadership at the automakers. Roger Smith left GM the same year California issued the ZEV mandate, and there appeared to be far more hostility than enthusiasm for electric cars among other auto executives. One of the most curious episodes in automotive history unfolded next, as car companies sought to undermine their own new products. First the car companies all refused to sell the electric cars. They would only lease them, and then only after a lengthy, intrusive application process. The electric car marketing and advertising were so inept that the poor quality could only have been purposeful—one television spot for the EV1 appeared to be promoting doomsday, not a cool new car.[11] And yet customers flocked to the paltry number of electric cars that trickled into showrooms starting in 1996. Celebrities such as the actors Mel Gibson, Tom Hanks, and Ed Begley snapped them up, and several prominent figures began to promote the idea of buying an electric car as an act of civic virtue. After Hanks raved about his EV1 on the television show *Late Night with David Letterman*, the perplexed host asked just what Hanks thought he was doing with an electric car. "I'm saving America, Dave. That's what I'm doing," Hanks replied. The studio audience applauded wildly.

Despite the consumers' excitement, as the deadline to increase production approached, the car manufacturers began complaining, after fielding only a few thousand, that their electrics were not practical, that American drivers would never accept the limited range, that the batteries were not good enough or durable enough, that they would go broke building vehicles nobody wanted. The carmakers minimized actual progress in battery technology that far exceeded their public admissions, particularly a new type of nickel battery invented by Ovonics—a company whose patents were bought by General Motors and whose batteries were suddenly no longer available to start-up companies working on new electric cars. The batteries had shown phenomenal promise, producing in one prototype sedan a range of 375 miles on a single charge.[12]

Yet, faced with corporate titans, teams of lawyers, and paid consultants, all insisting that a practical electric car could not be mass-produced, the political appointees and career bureaucrats of the air resources board blinked, first delaying the ZEV mandate at the automakers' request, and then, in 2003, killing the electric car program entirely, bowing to the car companies' eagerness to launch a new hydrogen fuel cell program. Instead of the originally mandated 150,000 nonpolluting electrics on the street in 2003, the car companies would have to deploy only 250 experimental hydrogen cars, with the number rising to only 25,000 by the year 2014. This was done over the objections of existing electric car leaseholders and a long line of scientists, including Andy Frank, who testified before the board that the carmakers were wrong about the batteries and wrong about hydrogen, and that it would be better if the ZEV program simply shifted a little to a "very low emissions vehicle" program—turning to plug-in hybrids rather than killing the electric car completely. His suggestions were dismissed, as was his (accurate) prediction that the carmakers would be back in a few years demanding more time for hydrogen, too.[13]

The electric cars were rounded up as their leases expired, carted off to the desert, and stacked in remote junkyards—just as the old trolleys had been piled fifty years earlier. Then the carmakers had those perfectly good, sought-after electric cars crushed and shredded into cubes of scrap. Only Toyota relented after repeated public protests and letter-writing campaigns, allowing some leaseholders to buy the cars they loved; and this one act of generosity showed that the car companies had lied about the durability and quality of the electric cars. Southern California Edison had bought a fleet of 320 of the electric RAV4-EVs, using them for heavy-duty work under harsh conditions and with frequent charging cycles. The original battery packs—the ones that were supposed to be too poor for the market and that would wear out too fast—were still in service in 2008, with 150,000 miles on them.

Ironically, the ZEV mandate was based on a mistake. The GM Impact that started it all was never intended by its designers as a model for a production car. Smith had just gotten carried away. Its designers had intended the Impact as a test bed for electric car technology—regenerative brakes, transmission, computerized controls—but that was supposed to be just step one. Once the kinks were worked out of the electrical systems, the EV1 was supposed to be upgraded to a plug-in hybrid.

"It would have been a great car," Andy Frank says with a sigh. He has a picture of the one prototype plug-in hybrid EV1 that was built, a bright red sports car that could travel up to eighty miles on electricity alone, and on longer trips could get the equivalent of eighty miles a gallon. "It never saw the light of day, but that type of car was the way we should have gone. GM could have sold thousands and thousands of them. California could have eased into the plug-in hybrid instead of pushing for the all-electric car, and many of the objections would have just gone away. Range isn't an issue then. The batteries don't have to be perfect, just good enough. There's a comfort zone. We'd be so much farther along."

Frank should know. He's the man GM hired to turn the EV1 into a plug-in hybrid—before it decided to kill the whole program and build Hummers instead.

Even as a teenager, Andy Frank loved to take apart old cars, scavenging old parts, then putting everything back together better than before. In the late 1940s and early 1950s, "better" meant faster. It meant more torque, more horsepower. At the time, the only way to improve fuel efficiency was to pull out unused parts of the car—the deadweight—to make it lighter.

Frank had emigrated with his family from China at age seven, just before America entered World War II. The family settled in Pasadena, and Frank grew up at the center of California's car culture. It was second nature for him and his high school friends to buy old cars, aged Model A's or Model T's, and juice them up for street racing—the hot rod craze. Once he put an immense Cadillac V-12 engine inside a 1936 Ford Phaeton that normally carried a motor only half that size and power. He would have needed a jet engine to make it any more powerful. "Well," he says with his easy laugh, "you have to have fun with the stuff."

His work today resembles his early exploits—he is still tearing apart cars, making them better, and calling it fun, although the primary goal nowadays is to shrink the gas engine to little more than a backup rather than make it roar and spit fire like a rocket. Since he started his automotive experiments in the 1970s, he has called his team of engineering students and volunteers "Team Fate," a reference to the 1965 classic slapstick film *The Great Race*, in which one of the main characters, Professor Fate, goes to any lengths to build a superior, gadget-filled supercar to win an impossible road race from New York to Paris. To this day, when he takes pictures of the team with their latest automotive handiwork on display, Frank assumes his Professor Fate persona, smiling broadly beneath an absurdly tall black top hat.

"Everyone loves Professor Frank," an engineering student, Andrew Shabashevich, observed after working all night to prepare Team Fate's latest entry in the Challenge X competition. This is one of several marquee alternative vehicle contests sponsored each year for major universities and research centers by the U.S. Department of Energy and automakers as tests for future technology. "I mean, we must, because the hours are just insane. Worth it, but insane."

Frank took a detour from cars after high school, earning undergraduate and master's degrees in mechanical engineering at Berkeley,
then working in the aerospace industry developing technology for the
Apollo moon mission spacecraft and the Minute Man Missile program. But after earning his PhD in electrical engineering, he turned
back to cars, initially as a young professor at the University of Wis-
consin. There, Team Fate got its start, with a goal that Frank still
maintains as his minimum for new designs: Build a car that gets 100
miles to the gallon and can go zero to sixty in ten seconds or less. He
believes this combination of economy and performance is key to winning consumer acceptance.

Frank built his first plug-in hybrid in 1972 for the Department
of Transportation's "Urban Car Contest"—the first national competition to pioneer a new type of car with high fuel economy and
low emissions. Frank's was the only hybrid entry; it capitalized on
and improved an idea from the early 1900s, when the first successful
hybrid, the Mixte, was produced and then abandoned, by the legendary car designer Ferdinand Porsche (whose first car was all-electric).
Frank had a completely different design, but he encountered the same
limitations Porsche had faced. The best lead-acid batteries were still
not up to the task, and modern nickel metal hydride and lithium-ion
batteries were still decades in the future. Even more problematic were
the control systems. The merging of the two separate electric and
gasoline power sources had to be continually adjusted, moving from
all-electric to various mixes of power, depending on driving conditions, speed, and the level of charge left in the batteries. Today, those
tasks are handled by silicon chips and software that make thousands
of small adjustments seamlessly; the driver does not even have to
think about them. In the early 1970s, before the age of small personal
computers, Frank had to cobble together mechanical solutions—
clever, but inefficient, demanding constant attention from the driver.
He shelved that research for the rest of the decade when the Department of Energy offered him grants to research mechanical flywheels

in vehicles as a means of storing energy—an interesting line of research that taught him much about miniaturization of components and the use of lightweight composite materials. By storing energy in the flywheel instead of a battery, Frank's flywheel hybrids increased fuel efficiency by about 50 percent, but the limit seemed to be about thirty-five miles a gallon, far short of his goal of 100 miles a gallon.

In 1985, Frank moved to the University of California–Davis, and started up the U.S. Department of Energy Center for Hybrid Electric Vehicle Research, where his work has been supported by Nissan, GM, Ford, and various government agencies as he produced one breakthrough vehicle after another with his team of students. One experimental vehicle set a world record for fuel efficiency in 1992 (3,300 miles per gallon). But his main contribution has been a series of consumer vehicles modified with modern battery packs tucked under the floorboards, scaled-down gas engines, and electric motors with a minimum of sixty-mile all-electric ranges. Frank holds thirty-five patents; these include his continuously variable transmission, which is crucial to his hybrid's efficiency; and his take on the entire plug-in hybrid car itself.

In Frank's vehicles, unlike all-electric cars, the range is not limited by battery charge, because on long trips, the small gas engine kicks in to keep the wheels turning—the battery is never allowed to drop below 20 percent charged. Even on those extended trips, when averaged out annually on the basis of typical American driving patterns, Frank's cars get 100 to 200 miles to the gallon with a gas engine about as big as that of a medium-size motorcycle—yet powering a full-size Ford Explorer SUV that goes just as fast as (or faster than) the factory original. Owning such a car means going to the gas station no more than four times a year. This was the modification Frank performed on the EV1 for General Motors before the program was killed—at around the time General Motors went in a radically different direction, buying out the Hummer line from the now defunct American Motors and closing down the EV1 manufacturing line. In vain, Frank pleaded with the California air resources board to move its mandate to plug-in hybrids, but the board, transfixed by hydrogen fuel cells, had been convinced by the auto industry that modern battery technology was no more adequate than the electric car technology of a century earlier.

There are additional potential benefits to widespread adoption of plug-in hybrids. Frank envisions using the cars as part of a home solar power system, storing the sun's energy in the cars, and using them to power homes during the late hours, when household electricity consumption is low and the batteries can easily handle the load. Workplace solar chargers in outdoor parking lots would turn hybrid cars into mobile power generation stations. Stationary battery packs in the garage of a home could make an even more capable system—Edison's suburban residence resurrected, with plug-in cars turning into power plants when parked.

"This is why we work so many hours on this project," Frank's student, Shabashevich, says. "We think this is the right path for the future of the world—the one where we actually *have* a future."

For years, Team Fate's cars have been a dominant presence in competitions sponsored by government and industry. The team's hybrid Mercury Sable—code-named Coulomb—was lauded as the best application of advanced technology in the Department of Energy's FutureCar competition in 1999; and the team's hybrid SUVs have achieved first, second, and third place in four FutureTruck challenges. Time and again, Frank's team has produced some of the most fuel-efficient and least polluting entries in the competitions, because this team is among the very few university research projects focused on plug-ins.

"This type of car, if mass-produced, can help us preserve our lifestyle yet wean ourselves from the oil addiction," Frank says—and he has been saying this at every auto conference he could attend for the past fifteen years. "If I can build these cars with college students and parts that a medium-size machine shop can produce, you would think the big car companies could do it, too, economically and on an assembly-line scale. But I've been beating my head against the wall for a long time."

Frank has had to fund his research with $60,000 of his own money to supplement his federal and industry grants—and provide food and living expenses for his loyal band of students. But the tide began to turn for his vision in 2006. Chelan County in Washington state, where three large hydroelectric dams on the Columbia River provide the cheapest electricity in America, sought a way to capitalize on its immense energy resources. Ron Johnson-Rodriguez, the county's director of economic development, attended a conference on alternative vehicles, looking for ideas. Like most Americans, he was only vaguely aware that California had tried and failed to pioneer electric cars. When he asked what had happened with that attempt, he got an earful about the controversy. And then he was told: Go see Andy Frank.

He did. Johnson-Rodriguez walked through the engine-strewn warehouse-lab at the University of California–Davis, drove one of Frank's cars, heard Frank's ideas for an electric grid in which plug-in vehicles provided energy storage, and was sold. First Chelan County, and then Washington state, mounted a campaign to acquire a fleet of plug-in hybrid vehicles to realize Frank's vision as a national test bed. And because car manufacturers had not yet produced a plug-in hybrid, Frank began helping the county create a center for converting conventional hybrids such as the Prius into plug-ins. Austin Electric in Texas, where wind power is being developed at a rapid pace, has begun a similar program with Frank's help.

Frank points out that there are more than 800 million cars on the road worldwide. That's enough cars to circle the world bumper to bumper—100 times. By 2050, the number of cars will triple. If the cars are not clean—indeed, if cars don't start becoming clean now, today—there will be a climatic Armageddon, he says.

The rising price of gasoline, consumers' declining interest in gas-guzzling SUVs, and the growing number of utility companies have finally forced Detroit's hand. All the major manufacturers have announced plug-in hybrids, although their specifications all fall far short of Frank's models; and Frank's company, Efficient Drivetrains Inc., is suddenly in great demand, particularly in Asia. But he doubts the sincerity of the major Detroit carmakers' commitment. After one presentation he made in 2007 to Ford's executives, he recalls one senior technology executive saying, "Well, Professor Frank, that's interesting. But what makes you think you have better technology than we have? . . . You're only one guy and we have the best technology in the world. We hire people from MIT and Cal Tech. Why are you better than those guys?"

Frank was flabbergasted. Ford and GM were on record as saying that battery technology was still too poor to get more than a forty-mile all-electric range, whereas Frank and his college kids had turned GM's own EV1 into an eighty-mile all-electric-range vehicle ten years earlier. "What was I supposed to say," he quipped later, "that he had hired a bunch of dummies?"

Such attitudes are dying a difficult, slow death, Frank says. Sooner or later, his work will transform the way people drive, because the results are undeniable. His cars are better than anything in any show-room anywhere, and the word is finally out: Andy Frank's cars can save the world.

"We will get there, finally," he says. "But the wasted time and opportunities are very depressing if I let myself think about them. We could be so much farther along. We could already be in a place where foreign oil would be irrelevant to us—we could *be* there, right now. Fear of change, and stubbornness, that has been the enemy."

can a malibu pool cleaner save the world?

Three years after the mandate for clean electric cars died a sense-less death, California took another stab at attacking smog and climate change when a movie star, pro-business Republican governor named Arnold Schwarzenegger ignored his staff's objections and turned to liberal Democrat and former pool cleaner Terry Tamminen, and gave him an epic assignment: Find a way to fix the mess.

The surprising part of it is, Tamminen may have actually found a way to do it. Through his efforts, the tools and determination to roll back greenhouse gas emissions may now be California's most im-portant export, spreading to other states and the private sector, with Tamminen in the self-described role as the "Johnny Appleseed of climate change."

"It really came down to not being able to wait any longer. We could not wait for a new president and a new Congress to wake up and save the planet," Tamminen says. "And what happened next just blows everyone's socks off, because even without federal leadership, by the time a new president is in the White House in 2009, a majority of Americans already will live in a state with a world-class climate change program, with measurable goals and a plan to achieve them."[1]

Arnold Schwarzenegger declared his candidacy for the governorship of California on the *Tonight Show with Jay Leno* on Wednesday, August 7, 2003. The celebrity venue suited the circuslike atmosphere of that election year, which included the recall of the sitting Democratic governor, Gray Davis, and a ballot listing 135 candidates who wanted his job. Most of these candidates were nonpoliticians; they included former child actor Gary Coleman; pornographer Larry Flynt (campaign slogan: "Vote for a Smut-Peddler Who Cares"); porn star Mary Carey; a comedian named Gallagher whose signature bit involved smashing watermelons with a mallet; and Angelyne, who become a minor Hollywood celebrity by putting up billboards of herself throughout Los Angeles that pretended she was already a celebrity.

Schwarzenegger, at the time best known for his role as a robotic assassin in the *Terminator* films, was immediately declared the sensible choice and the front-runner. The weekend after his announcement, he and his family flew to Massachusetts for a visit with his in-laws—the Kennedy family. Politics is the Kennedys' lifeblood, and he was peppered with questions about his platform, his policies, his plans, his staff, and his campaign. At that point he could offer few specifics beyond the conviction that someone from outside Sacramento with a businessman's eye and a nonpolitical perspective—this is how he defined himself at the time—could bring fresh ideas to the task of fixing California's considerable problems. One of the areas he intended to tackle was the environment, he said, because the dirty ocean, beaches, and air in Southern California troubled him. He needed help to translate his concerns into policies, though. As it happened, there was an expert in such matters standing right next to him. So Schwarzenegger asked his wife's cousin, the environmental activist Robert F. Kennedy Jr., who was the chief attorney for the antipollution group Hudson Riverkeeper and a senior attorney for the Natural Resources Defense Council, about whom he might approach. Kennedy told him, "Talk to Terry."

Terry Tamminen's varied résumé appealed to the multi-careered Schwarzenegger, an Austrian immigrant who has been an American bodybuilding champion, an action movie star, a real estate investor, a restaurateur, a "Global Torch Bearer" for the Special Olympics, and then, in 2003, a gubernatorial candidate. Tamminen's background was every bit as diverse. Born in Milwaukee, he had grown up partly in Australia, where his family had a business breeding tropical fish, and partly in Southern California, where he had been inspired to become a scuba diver at age twelve by the old action television series, *Sea Hunt*. He traveled and worked in Europe, came home for college in California, left to convert apartments to condos in Florida, ran a sheep ranch owned by his boss at the condo company, and finally moved back to California, where in 1976 he bought a half-share of a pool cleaning business based in the swimming pool capital of the world, Malibu. He lived modestly in a trailer near the coast, but this gregarious jack-of-all-trades with the round, cherubic face and an emphasis on eco-friendly water-quality practices, gradually built up the business's customer base tenfold. He catered to celebrities in Beverly Hills and Malibu—Madonna, Johnny Carson, Dustin Hoffman, Dick Clark, David Letterman, Kareem Abdul-Jabbar, Bruce Willis, Sting, Diana Ross, and Barbra Streisand, along with many agents, producers, and other movers in Hollywood. He wrote several books on pool maintenance, including *The Ultimate Pool Maintenance Manual*, styling himself the "Malibu Pool Man to the Stars." Then he wrote a stage play about Shakespeare as the bard entered retirement, which he has performed at community theaters and colleges nationwide.

In his forties, still feeling unfulfilled, and experiencing what he calls his midlife crisis, Tamminen sold his share of the pool business and started an environmental group, Santa Monica BayKeeper, in 1993. It was funded by the late former president of the Disney Company, Frank Wells, and modeled after Kennedy's Hudson Riverkeeper group. Tamminen earned his certification from the Coast Guard as a ship captain, and he and a cadre of volunteers would patrol the bay as a citizen pollution police force, reporting violators and deterring dumpers—and Tamminen's education as an environmentalist began. He took a crash course taught by Kennedy and soon realized that he had at last found his great passion in life. His wife would get up each morning and put on her business suit while Tamminen would pull on a wet suit to begin his workday—a dream job for the kid who grew up watching *Sea Hunt* and the popular 1970s television documentaries by the French explorer and ecologist Jacques Cousteau. Later Tamminen also worked to set up and direct a foundation, Environment Now, to provide grants to similar WaterKeeper programs throughout California, but he always returned to the boat. "Glad to see you finally cleaning up the big pool," Tamminen recalls a former customer, Dustin Hoffman, quipping the next time they met.

As with other eco barons, Terry Tamminen's mission in life began with an affinity for nature—his passion for boating and diving and his desire, his need, to be outdoors for significant portions of his life. Kieran Suckling argues that Americans have built a culture so distanced from any real experience of nature—most people, he says, watch it on television more than they actually touch, smell, or feel it—that society can casually lay waste to the natural world and most people never perceive it. Distance makes desecration possible, because for most Americans, the experience of nature comes through a television documentary. They think that the cut and recut forests of America, with undergrowth abnormally thick from years of suppressing the normal cycles of fire and regeneration, are natural. They do not know what a real forest should look like—the kind of forest that greeted the pilgrims and the explorers, that Thoreau and John Muir loved and watched vanish. If Suckling is correct about this distance from and desensitization to nature, then it is no coincidence that the Tompkinses, Sucklings, Galvins, Quimbys, and Tamminens of the world, people who are most happy in the woods, on the water, or clinging to a mountainside, are able to experience and share environmental insights that elude others, and that the world desperately needs. Tamminen found his place and insight on the bay, but it also set the stage for his next career.

And then Arnold Schwarzenegger called.

Schwarzenegger was so eager to put the green part of his campaign in place that he phoned while still at the Kennedy compound in Massachusetts. Would Tamminen meet with him the following week?

A few days later, they were sitting in Schwarzenegger's office in Santa Monica, California, drawing up a list of campaign pledges for the environment—ambitious proposals on fuel efficiency, tailpipe emissions, alternative energy, and a "hydrogen highway" for fuel cell cars; new rules on water quality and pollution; and a vow to develop quickly a serious statewide plan to combat global warming. When Tamminen's friends and colleagues heard of this unlikely alliance— most of Schwarzenegger's advisers were extremely conservative staffers left over from the last Republican governor, Pete Wilson—they thought he had lost his mind and were quick to say so. Tamminen was a lifelong liberal Democrat. He had been a fund-raiser for Al Gore's 2000 run for the White House, and Tamminen's Democratic friends considered any association with "the party of George Bush" as bordering on treason. Tamminen, however, had been pleasantly surprised during his meeting with Schwarzenegger, who, he felt, demonstrated a solid understanding of climate change and acknowledged it as a threat to the world and especially to the coastal state of California. In Tamminen's view, this attitude put Schwarzenegger far

ahead of where most American politicians and most citizens were in
2003, and certainly ahead of the California Republican Party, which
was dominated by the far right.

"First of all, this guy might be your next governor," Tamminen
argued, "so let's make sure we fill his head with the right ideas."

No one was buying it, though. "I took a lot of crap," Tamminen
would later recall. "But I saw the opportunity, possibly, to shape not
just California's environmental future, but by virtue of California's
tendency to lead the rest of the nation, to have a tremendous impact."

It was all academic, however, Tamminen thought at first. Mainstream politicians from both parties had entered the race, and
Schwarzenegger's candidacy had come to seem quixotic; Tamminen
figured that his own main contribution would be to put forward a
plan that other candidates would be forced to match,[2] and after the
election, he could return to his old life. But as Election Day neared, it
suddenly dawned on Tamminen and his comrades that Schwarzenegger might actually win. At this point, Tamminen made it clear that
he had no interest in a career in government, or in leaving his boat
and the glorious coastal climate of Santa Monica to endure the heat
of inland Sacramento. He had laid out what would be the most far-
reaching and groundbreaking environmental program in the state's—
and the nation's—history. His job was done.

A couple months later, he was sworn in as the newly elected Governor Schwarzenegger's secretary of the California Environmental
Protection Agency, charged with making their campaign proposals a
reality. "What can I say?" he shrugs. "I couldn't say no to Arnold."

At the outset, the worst fears of Tamminen's friends seemed to
be realized. In his first years in office, Schwarzenegger the moderate
candidate appeared to morph into a hard-line conservative governor,
who engaged in a running battle with teachers' and nurses' unions
over budget cuts he had promised not to make, having regular show-
downs with legislators (whom he labeled "girlie men," borrowing a

line from an old television skit on *Saturday Night Live* that had originally mocked Schwarzenegger), and drawing praise from conservative Republicans nationwide. Conservatives began openly discussing
the possibility that Schwarzenegger might run for the presidency,
though this would require a constitutional amendment to lift the restriction, imposed by the Founders, that the president be a native-
born American.

Tamminen hunkered down and continued working on the environmental action plan, confident that Schwarzenegger's rightward tilt
was more media story line than reality, because his own marching
orders and dealings with Schwarzenegger hadn't changed. And he
made some substantial progress that first full year in office: creation
of a 25-million-acre Sierra Nevada Conservancy to protect undevel-
oped land in twenty-three counties stretching from central California to the Oregon border; ocean protection regulations that banned
dumping by cruise ships; an initiative to subsidize the installation of
solar panels on 1 million roofs in California; $150 million in perma-
nent annual funding for the Carl Moyer Program, which pays for the
scrapping of dirty diesel buses and trucks and replaces them with vehicles that run on natural gas or other clean fuels; and the creation of
the promised "hydrogen highway" program, with an ambitious goal
of building 200 fueling stations by 2010 for clean hydrogen cars that,
at the time, existed only in theory. (The overly optimistic embrace
of the expensive and still unsuccessful hydrogen technology was the
most glaring flaw in the Schwarzenegger and Tamminen approach to
global warming, but environmental activists were so thrilled by the
rest of the plan that criticism was muted.)

Despite a successful start at fulfilling a substantial number of
environmental campaign promises, Tamminen's work that year was
marginalized by some of Schwarzenegger's other staffers, who considered him too liberal. There was little effort to publicize or generate
excitement about Cal-EPA's accomplishments.

But the dynamics changed the following year. Schwarzenegger's conservative policies, confrontational tactics, and bashing of teachers and nurses proved deeply unpopular, his rating in the polls plummeted, and his attempt to go around the Democratic-controlled legislature with a series of ballot provisions to enact his conservative agenda ended in disaster, as voters rejected all his key proposals. After the vote, he did something few politicians do, and it saved his administration: He publicly apologized, and his policies took a markedly more pragmatic and moderate turn as he distanced himself from both President Bush and his own party's state leaders. The talk about Arnold in the White House and a constitutional amendment evaporated, to be replaced by serious talk of global warming that posed a direct challenge to his party and the president.

"I say the debate is over. We know the science, we see the threat, and we know the time for action is now," Schwarzenegger announced in June 2005, unveiling his new state plan to tackle climate change. Tamminen and his staff had crafted an aggressive, measurable program to reduce greenhouse gases, increase conservation efforts, and increase the use of alternative energy. This program would make California a world leader in the race to blunt the worst of global warming, with an ultimate goal of lowering greenhouse gas emissions so much that by the year 2050, they would be 80 percent lower than in 1990.

It had taken more than a year to study the problem, inventory the state's greenhouse gas emitters, and develop a plan. The task was slowed because when Tamminen arrived there was not a single climate scientist at Cal-EPA. Within two years, he had 200 people in state government working on the problem.

This was in sharp contrast to national climate policy under the Bush administration, which lagged behind all other developed nations in responding to climate change and sought to deal with rising energy costs by endlessly proposing drilling in the protected Arctic National Wildlife Refuge and other ecologically sensitive regions on and offshore. Even the most optimistic estimates about such drilling predicted a level of oil production that would be only drops in the bucket compared with the potential savings a genuine pursuit of conservation and alternative energy sources could yield.

For any state, even one as large as California, to attempt to lead the nation forward on global warming was a radical departure. In the past, it had always been the federal government, after much delay and prodding from activists, that dragged reluctant states into a more enlightened future: Teddy Roosevelt's forests and parks conservation programs; the equal employment, fair housing, voting rights, and desegregation laws and court decisions of the 1950s and 1960s; and, more recently, the landmark clean air, clean water, and endangered-species initiatives of the 1970s. Energy efficiency and fossil fuel emissions were also supposed to be a federal prerogative, but federal leadership had mostly languished since Jimmy Carter left office in 1981. Carter has not been revered as a president (he is admired more for his charitable and human rights work since leaving office), but history may actually be kinder to Carter than to his successor, particularly on the subject of energy and the environment. Had Carter remained in office for a second term rather than being ousted by Ronald Reagan, the nation (and the world) very likely would be facing a less severe, less imminent climate catastrophe and a milder oil addiction problem in the twenty-first century. Carter had a plan in place to begin fixing the mess, and many years later it provided a starting point for Tamminen. Reagan's actions, on the other hand, made matters worse, and provided an example of what not to do about environmental and energy problems.

Carter had made renewable solar energy a central part of his program, putting the founder of Earth Day, Denis Hayes, in charge of a federal Solar Energy Research Institute. Hayes brought an activist's passion and single-mindedness to the bureaucratic Department of

Energy, just as Tamminen would do in Sacramento a quarter century later. Hayes commissioned a one-year study by leading scientists, who concluded in 1979 that existing technology could be used to
convert 28 percent of the nation's energy use to nonpolluting renewable sources, primarily solar, but also wind and geothermal. All that
would be required was a system of incentives and disincentives—the
government had to make it more profitable for utilities to produce
renewable energy and more expensive to produce dirty power dependent on fossil fuels, and then provide similar incentives to consumers. Likewise, efficient low-polluting or nonpolluting cars would be
rewarded with tax breaks and other incentives, whereas gas-guzzling
smog machines would be penalized.

It was only fair to set up such a structure, environmentalists
argued, as it would be no worse a violation of free-market principles
than had previously existed: oil, nuclear power, and coal had been
subsidized for decades. The tens of billions of dollars of annual tax
breaks were not the only subsidies to Big Oil and Big Auto—these
industries had also been held harmless for the enormous health costs
from toxic fossil fuel emissions. Normally, it is a fundamental legal
principle that if a product causes harm by design, the producer ought
to be responsible. Rates of cancer and lung diseases are higher in
smoggy cities and higher still in neighborhoods with heavily traf-
ficked highways cutting through them, but neither the automobile
makers nor the oil companies have ever had to pay a cent to cover the
medical costs (or funeral expenses) their products caused. Instead,
consumers and, ultimately, taxpayers have always footed that bill—a
subsidy worth trillions of dollars to those industries. Industry had, in
effect, been given incentives to make the earth toxic. Now the time
had come to turn that situation on its head and make it profitable for
the oil and automotive industries to clean up their act, and unprofit-
able to fail to do so. As Terry Tamminen would say many years later,
"They have much to atone for."

The solar program Carter launched was not primarily a matter of reducing pollution, and global warming was then merely an obscure concern of a few climate scientists who had no impact on public policy. Carter's interest in solar energy was primarily a matter of national and economic security—a rationale that remains valid in the twenty-first century. It stemmed from two oil crises, in 1973 and 1979. The first crisis came in the form of an Arab oil embargo aimed against Israel's allies during the Yom Kippur War. It led to rationing, long lines at gas stations, and spiraling fuel prices during Richard Nixon's presidency. Congress responded by adopting fuel efficiency laws requiring automakers to gradually improve fuel economy so that new cars would consume less gas—the now famous Corporate Average Fuel Economy (CAFE) standards.[3] The second oil crisis came in the form of severe shortages resulting from a war between Iran and Iraq, and from the Iranian revolution—during which fifty-five members of the U.S. embassy staff in Tehran were held hostage for 444 days, destroying President Carter's chances for reelection. Oil prices shot up during the crisis of 1979 to thirty-nine dollars a barrel (ninety-eight in 2008 dollars, a price record that stood until March 2008). This second oil crisis persuaded President Carter that national security depended on immediately launching an urgent program of conservation, solar energy research, and incentives to stimulate the adoption of alternative energy sources. He told the nation that America needed to free itself from dependence on foreign oil suppliers and, appearing in a sweater on national television, urged Americans to lower their thermostats and dress more warmly in winter instead of cranking up the heat. The solar panels he put up on the White House were intended to be a symbol of his and the country's commitment to breaking the oil addiction, and thanks to tax credits and other incentives, the solar power industry took off as never before.

But it ground to a halt under Reagan. Solar research funding was cut; most of the staff was let go. The White House solar panels were trashed. Hayes's study and its ambitious target of 28 percent alternative energy for the nation were shelved and forgotten, as Reagan

declared in his first inaugural address, "Government is not the solution to our problem; government *is* the problem." He spoke then
of recapturing America's past glories, which to him meant pursuing not conservation or solar energy, but more privatization of the
oil, gas, and coal deposits on public lands, so that Americans need
not alter their driving habits, their thermostats, or their addiction to
oil. Reagan and Congress even halted regulations that would have
required continuing improvements in automobile fuel efficiency. For
the next quarter century, the requirements for miles per gallon remained locked in where Carter had left them.

The solar energy incentives were withdrawn without any tapering
off—a devastating blow to the nation's fledgling solar industry. The
same happened with wind energy. Carter's tax credits allowed the
U.S. wind-power industry to grow and take world leadership: American companies controlled 90 percent of the world market by the early
1980s. Then the tax subsidies were cut and the industry withered; the
United States now has only 16 percent of the market, and countries
with government tax breaks have taken the lead. Promising research
on geothermal power has also languished compared with research by
other nations. Instead, Reagan launched a very different research program, the Strategic Defense Initiative, a missile shield project begun
in 1983. It is the most expensive weapons system in history; more
than $100 billion was spent over twenty-five years, and the second
President Bush appropriated more for it than ever before—$11 bil-
lion a year in his last years in office. After twenty-six years, it still has
never been shown to work. One year of funding for "Star Wars," as
this program has been called, represents more than the government
has spent on solar energy research in twenty-five years—even though
solar energy is a technology that could immediately improve national
security and help address concerns about climate change. Had the
nation continued as Carter intended—gradually converting to clean,
emission-free solar power and giving industry the incentives it needed
to pursue alternative energy seriously—America could have become

a world leader in solar power within a decade. Instead, the nation is playing catch-up and generating less than 1 percent of its power from the sun—while importing technology from the world leaders in solar energy in Europe and Israel.

By 1997, when an international agreement to battle climate change, the Kyoto Protocol,[4] was negotiated in Japan, America was ill-equipped to participate in this new framework for reducing greenhouse gas emissions. Kyoto covered 170 countries, and the industrial nations were expected to adopt binding reductions in greenhouse gases phased in during the coming decades. Developing nations had voluntary goals under the treaty; these nations included China, which has become a major emitter of greenhouse gases as its economy rapidly expanded. Alone among the developed nations, America has refused to ratify the treaty (Australia, the only other holdout, ratified it in 2007). President Clinton signed the treaty but, faced with nearly unanimous opposition in the U.S. Senate, he never sought the required ratification. President Bush said he would not agree to Kyoto because it put America at an economic disadvantage and because China was not part of the mandatory caps. The result is that the United States has had no meaningful, sustained mechanism in place for reducing greenhouse gas emissions. And because America has long led the world in emitting greenhouse gases, the entire world was imperiled by its failure to act. (In 2007 China finally surpassed the United States in overall greenhouse emissions, thanks in large part to the construction of numerous dirty coal-fired electrical generating plants in China. But America remains by far the per capita leader in this dubious competition to warm the earth.)

So it fell to California to take the step that neither Clinton nor Bush had undertaken.[5] Tamminen started with what he considered a commonsense approach, seemingly obvious—though few others in positions of power in government were yet endorsing it. He started with the science of climate change, not the politics: The question was what could be done scientifically to reduce greenhouses gases, not what would satisfy a political constituency. Next there had to be a complete inventory of where the emissions were coming from—utilities, transportation, agriculture, construction, manufacturing, mining—and in what proportions: No workable plan could be made without a thorough inventory of California's greenhouse gas sources—it would be like trying to build a skyscraper without knowing how much iron, glass, and concrete were on hand. Those steps provided the amount of reduction that could be sought, and the targets. Then it was possible for Tamminen to sit down with the governor to consider the menu of options available to reduce greenhouse gas emissions, based on science and the experience of other countries, devising a plan that fit California's needs.

Writing that plan was not the breakthrough, however. For years, scientific organizations, government committees, and activist groups have been coming up with national and regional plans to grapple with global warming, some little more than political posturing, some deeply flawed, some visionary, others practical. What was different here was the will to act on Tamminen's ideas at the highest level. The action plan Tamminen and Schwarzenegger settled on exceeded the goals set by Kyoto, in part by applying a cap and trade system through which businesses received credits for reducing greenhouse gases below the maximum emissions cap. They could then trade or sell some of those credits, in the form of carbon offsets, to other companies that had not sufficiently reduced their carbon footprints on their own.

Development in cities and counties would have to take greenhouse gas emissions into account—in construction methods, in choice of locations, and in provisions for mass transit and solar energy, just as the Center for Biological Diversity has advocated. The same conditions would apply to new industry and businesses—the plan provided incentives for being clean, and consequences for continuing with business as usual. Californians would have to be told—over and over, according to Tamminen—that the state faces a unique threat of global warming and that this threat requires big changes. The alternative to change would be to experience mass extinctions of up to 50 percent of the state's unique plants, including the majestic Redwoods, in the coming decades, along with coastal flooding as the sea ice continues to melt, and along with drought everywhere else. "People think it will be too disruptive to change what we drive or how we build," Tamminen says. "And we're here to tell them that's nothing compared to what will happen to us if we *don't* clean up our act. Now, that will be disruption, on an epic scale."

Tamminen's plan would also mark a return to California's aggressive mandate for low-emission cars—automakers worldwide would be required to reduce their vehicle greenhouse gas emissions by 30 percent by 2016, or stop selling cars in the nation's biggest car market. This time Tamminen felt that all technologies should be brought to the table—electric, hybrids, hydrogen. Anything was fine, so long as the greenhouse gases went down the requisite amount. This approach, it was thought, could avoid the earlier, disastrous ZEV mandate and its thirteen wasted years. He envisioned more of partnership with the car, utility, and other industries, rather than the confrontations of the past, although he knew automaker lawsuits over emissions regulations would be inevitable.

Alternatives to fossil fuel energy generation would have to play an important role in the plan; these included both small-scale solar power systems on individual homes and businesses (through the state's Million Solar Roofs Plan) and large-scale solar power plants. The plan envisioned massive solar generating complexes in the Mojave Desert, one of the sunniest places on earth, where several such plants, using solar-heated steam turbines to drive generators, were already on the drawing boards.

Tamminen did not structure the whole plan around any particular technology. Instead, he chose a "renewable portfolio standard"—a smorgasbord approach that used both mandates and market-based methods, an embodiment of his often repeated refrain: "There are no silver bullets for climate change, only silver buckshot. We have to cast a wide net." Whatever silver buckshot the state ended up using, there was nothing ambiguous about the portfolio's goal: it required 20 percent of the energy used in California to come from renewables by 2010, and 30 percent by 2020. At the time the plan was designed, renewables accounted for about 4 percent of the energy generated, so this was an ambitious goal. It was meant to have an immediate impact, and if it was adopted nationally it would spur the kind of research on solar and renewable energy that Carter had attempted to start and Reagan had quashed.

Once completed, the Tamminen-Schwarzenegger plan became the basis of California's Global Warming Solutions Act, which passed the state legislature on a party-line vote, with all but one member of the governor's party voting against it. The legislation made the aggressive reductions in greenhouse gas emissions legally binding—something no other state had done.

California could do this because of its unique legal position among states when it came to environmental regulation. Because California had been beset by smog since the 1940s, it had undertaken pollution controls and regulation of auto emissions before the U.S. Environmental Protection Agency was created in 1970. So Congress—over fervent opposition from industry—decided to give California free rein to choose between federal regulation and its own, and California's could be more stringent. No other state had that power—although other states had the option of following California's more stringent standards rather than Washington's.

This is why California was able to lead the whole nation in environmental regulations. And California really had led the nation since the 1960s: It was first to regulate smog-causing nitrogen-oxide emissions, first to set standards for diesel truck emissions, first to require catalytic converters to absorb tailpipe emissions in cars, first to introduce unleaded gasoline, and first to require automakers to produce a certain percentage of low-emission and zero-emission vehicles. Oil and automotive companies fought all these initiatives and lost; most of the measures were later adopted by other states or the federal government. California had also made its household electrical consumption twice as efficient by imposing a rate structure that rewarded conservation and made it profitable for utilities to sell less energy rather than always trying to sell more. Normally, pushing conservation hurts the bottom line for a utility, but in California, it increased profitability. Because California would be the sixteenth-largest greenhouse gas emitter in the world if it were a separate nation, any large-scale conservation measures it takes have a global impact.

The states of Washington and Arizona followed California's lead and adopted similar initiatives to deal with global warming, and Tamminen served as an informal consultant. Then, during a meeting of the Western Governors' Association, Schwarzenegger, Tamminen, Governor Janet Napolitano of Arizona, and Governor Christine Gregoire of Washington discussed their fear that three states going it alone would see businesses leave for places with no such regulations on greenhouse gas emissions. And that outcome could doom their efforts. Without a national policy, they needed more states on board, so that business would simply accept the change rather than move, and perhaps eventually recognize the program as a potential opportunity to save money, and even profit by selling carbon credits.

Tamminen saw this as his opportunity to get out of government, yet still continue his work. He offered to become the "Johnny Appleseed" of climate change—he would export California's global warming program to other states, acting as Schwarzenegger's emissary and consultant.

Within a year, thirteen states had climate action plans in place, most based at least in part on California's Global Warming Solutions Act. Another fourteen states had plans in development, and seven more had greenhouse gas assessments under way in order to write plans of their own. Agreements were also signed with British Columbia and the United Kingdom for participating in an international cap and trade program to reduce greenhouse gas emissions—California was bypassing the Bush administration on the international climate policy stage just as it was doing at home. Tamminen had made a particularly hard push in the southeastern United States, a region that lagged behind the rest of the nation in responding to global warming. He made a detailed presentation on climate science to the newly elected Republican governor of Florida, Charlie Crist. Crist, a former state attorney general, had not run on an environmental platform, but Tamminen helped convince him that Florida's long coastlines, its extensive development at or near sea level, and its vulnerability to hurricanes left it uniquely imperiled by the onset of global warming. There was evidence that a pattern of heavier hurricanes and tropical storms—like the pattern of drought and heat afflicting California and the West—had been strengthened if not actually caused by global warming. Crist decided to make an aggressive climate action plan a top priority for his new administration, with Tamminen as a consultant. In July 2007, just seven months after taking office, Crist announced a climate change program that matched California's in its reach and ambition, including the same overall goal: By the year 2050, both states vowed to reduce greenhouse gas emissions by 80 percent relative to the level of emissions that existed in 1990.

With two large and politically pivotal states—California and Florida, east and west coasts, both with Republican governors— Tamminen declared that the country had reached a tipping point in the battle against global warming, despite inaction and obstruction at the federal level. A majority of Americans were residing in states with plans to address global warming either in place or fast approaching, and those states would be tied to one another's carbon-trading plans, as well as to Europe's and Canada's—a truly global solution.

Tamminen had pushed George Bush and the federal government to the sidelines in the battle against global warming—except in their accustomed role as obstacles to action. Although the Bush administration had always championed state and local solutions to problems, the White House now attempted to block important portions of California's Global Warming Solutions Act. In a move reminiscent of its flouting of the Endangered Species Act, the administration simply refused to follow the law granting California special authority to create its own auto emissions regulations. To invoke this special authority, California had to request a waiver from the U.S. Environmental Protection Agency. This was strictly a legal formality, as federal law does not give discretion to the EPA to withhold a waiver—the law says the EPA "shall" give the waiver so long as the state's regulations are at least as protective as the federal version. And there was no question that California's emissions standards were the stricter by far. Yet Stephen Johnson, Bush's appointee to head the EPA, denied the waiver in 2008 after months of delay. He argued that global warming had to be addressed only on the federal level because greenhouse gas emissions harmed the whole nation, not just Californians. His explanation studiously ignored the fact that there was no similar comprehensive effort to address global warming on the federal level—California, the Center for Biological Diversity, and other states and groups had just successfully sued the EPA because Johnson had claimed the federal government had no power to regulate greenhouse gases at all, a position soundly rejected by the U.S. Supreme Court. In thirty-seven years, no waiver request from California had ever been denied. Under the first President Bush alone, the EPA granted nine waivers in four years, and Johnson's scientific and legal staff told him there were no legitimate grounds to deny this one. But he was adamant: The waiver request was rejected.

As California took the Bush administration to court over this denial, Tamminen remained sanguine. The delay was unfortunate— reducing auto emissions was essential to his plan—but he knew that the law and the science were on California's side, and, more important, so was the calendar.

When the Bush administration denied California the right to enforce its own global warming battle plan, there were only three candidates left in the race to be the next president—and all three had said they would approve the waiver. So either way, through the courts or through ballots, come 2009, the obstruction would be gone, and Tamminen's plan for California to lead the nation in a war against climate change could begin in earnest.

the turtle lady

Terry Tamminen found a way to strike a blow for environmental protection through collaborating with government and industry, but there are times when saving a piece of the natural world can be accomplished only through a more traditional approach: protest and conflict with the powers that be. So it was with Carole Allen and her quarter-century battle to bring the imperiled Kemp's ridley sea turtle back from the brink of extinction. She marshaled an army of schoolchildren to demand protections for an endangered species. She took on the powerful Texas Gulf Coast shrimping industry, which had long resisted reform. And she belied conventional wisdom that individuals are powerless in the face of monied entrenched interests.

As a result, the graceful Kemp's ridley, once the most endangered sea turtle in the world, is making a strong comeback—a testament to the fact that the measure of an eco baron is not a matter of wealth or fame, and that one committed person can have an outsize impact.

"I'm no millionaire, that's for sure," Allen chuckles. "But I do know right from wrong, and what was happening to these wonderful creatures, who have been around since the age of dinosaurs, most certainly was wrong."

Allen has been fascinated by turtles since age six, when she was growing up in Illinois and had a pet red-eared slider freshwater turtle she tried to cart around everywhere. Today her home north of Houston is something of a turtle shrine—with turtle statues, plush turtles, and turtle pictures.

She and her husband, Bill, a geologist, moved to Texas in the 1970s so he could work in the oil industry there. When he became ill with a heart condition, Allen decided to return to college so she could earn a degree and find work that would support the family, which by then included a young daughter. She received her degree in journalism just before Bill died, and Allen went to work for the juvenile probation department to support herself and her daughter.

In 1978, she read about a new experiment being conducted to restore a breeding population of Kemp's ridley sea turtles to the Texas Gulf Coast. The species is named for Richard Kemp, a fisherman in Florida who was interested in natural science and who recognized the turtle as a unique species. He submitted a specimen to a local university in 1906. The Kemp's ridley, once plentiful in the region, is a seafaring, air-breathing marine reptile with a heart-shaped shell. It grows to about eighty to 100 pounds, making it the smallest of the seven sea turtle species left in existence, all of them listed as endangered since the passage of the Endangered Species Act in 1973.[1] They live 100 years or more, and to breed and lay eggs they return to the same stretch of sand where they were born, thanks to their ability to sense the earth's magnetic field and use it to navigate for thousands of miles—nature's own Global Positioning System. In Texas, these turtles had been killed for their meat, shells, and eggs and, as unintended "by-catch," they also had been killed in great numbers by the busy gulf shrimping fleet. The shrimp vessels dragged long nets that often scooped up the turtles, which then died, unable to reach the surface to breathe. If they were hauled up still alive, the shrimpers might throw them back into the water to take their chances with the nets again, or just as often simply kill them as pests or for a meal.

Their last known mass breeding place, a few hundred miles south of Brownsville, Texas, was a beach in Mexico, called Rancho Nuevo. In 1947, about 40,000 of the turtles were filmed by a Mexican rancher as they emerged from the surf in a mile-long procession—*la arribada de las tortugas*, the arrival of the turtles—for a single day of mass nesting and egg laying in their ancestral sands. Poachers had a field day there, too, following right behind the stately tan and green tortoises, greedily scooping up their precious eggs to consume or sell at a tidy profit: There is a myth that the eggs contain a powerful aphrodisiac. By the late 1970s, the *arribada* numbered no more than 200 in a single day. The world population had declined to 3 percent of its original numbers, so this species was the most endangered turtle on earth. And since it had only one main nesting place left in the world, one ill-timed hurricane during egg-laying season could wipe out the species. The Kemp's ridley would soon be extinct, scientists feared. Some thought it was already too late to save the species.

The federally funded program Allen heard about, called Operation Head Start, was intended to restore a breeding population of the turtles to south Texas and thereby give the species a fighting chance at survival. Every year, Mexico—which had made harming the turtles or their eggs a federal crime—would donate 2,000 Kemp's ridley eggs (each turtle lays about 100 eggs at a time, so the surviving Mexican population of about 500 breeding females could spare that many). The National Marine Fisheries Service would fly the eggs to South Padre Island in south Texas to incubate in the warm sand until they hatched, so the hatchlings would form an "imprint" with the South Padre beach, and if all went well, return there to breed when they matured years later. After the imprinting, the turtles were brought to Galveston, where the fisheries service ran the nation's only marine sea turtle research and rehabilitation lab, and they were allowed to mature for ten months to give them a better chance at survival in the wild. Newborns generally had a high mortality rate; as the little turtles floated wherever the currents carried them, all along the Gulf and Atlantic coasts, as far north as Canada, they were eaten by just about anything that flies or swims. As they matured, they would return to shallow coastal waters, primarily in the Gulf of Mexico. The shells of the Head Start turtles were marked so that if they returned to their new nesting area, the researchers would know.

Allen was enchanted by the juvenile turtles and the program aimed at bringing them back. She visited the turtle hatchery in Galveston, Texas, in 1978, and three years later she talked the principal at her daughter's grade school into arranging a class outing there. Two hundred students made the seventy-mile trip one Saturday to see the freshly hatched babies, still in buckets of sea water after being plucked from the sand, black at that stage of life and only half the size of a child's hand. While the children learned about a turtle's life at sea, its urge to return to the sand on which it hatched, and the many threats that are making the Kemp's ridley ever more rare, Allen heard from the hatchery director that the new Reagan administration was cutting funding for the program. When she shared this with the elementary school kids, they immediately asked how they could help. The lab director pulled Allen aside and said pointedly, "What these turtles really need is a constituency." Allen translated for the kids: "If these turtles are going to make it, they need some friends."

Out of that, a constituency was born—elementary school children. Allen suggested a community awareness and fund-raising campaign to obtain feed for the growing turtles. The kids, inspired by the turtles' heart-shaped shells, suggested that the campaign be given the acronym HEART, for "Help Endangered Animals—Ridley Turtles," and the name stuck. It turned out that four dollars was enough to buy a supply of Purina Turtle Chow for one hatchling, and every kid who brought in donations was rewarded with a red heart cutout placed in a little cardboard turtle house. Allen worked as a volunteer in the classroom and then in the whole school, handling the donations and making and doling out the turtle houses. After one of the Houston newspapers picked up the story, Allen started getting calls and letters from around the state and the country, and soon dozens of schools and thousands of kids were raising money to save the sea turtles, and sending penciled letters to Congress and the White House, pleading for more government protection for the turtles. The program became both a fund-raising tool and a teaching opportunity for elementary schools, whose pupils learned about sea turtles, extinction, and the Endangered Species Act. In 1981 HEART became a nonprofit and a nearly full-time volunteer pursuit for Allen, in addition to her full-time paying job at the probation department. She and a group of other volunteers started making presentations to classrooms and civic clubs on the problem of the sea turtles' possible extinction, drumming up support for Head Start and for the turtles, which were among the earliest listings under the Endangered Species Act, yet had little or no protection and no critical habitat.

Every year in May, when the hatchlings had grown to the size of a laptop computer, Allen and a group of student volunteers—a few of them had started with HEART in elementary school and were still taking part in college—would go out with the marine fisheries service and release the juvenile turtles in the warm gulf waters. Gently tossing the turtles into the ocean had been a sort of an annual rite, a source of hope and inspiration for the volunteers. "This is what all the work and the fund-raising and the lobbying is for, this moment," Allen exulted after she had the honor of releasing the final turtle of 1987. She did not know it would be the last time for this yearly ritual.

The danger to the Kemp's ridley sea turtle was no longer from egg poachers—there were no more mile-long processions to raid, and the Mexican government sent armed soldiers to protect the few straggling breeding tortoises that clambered out of the water at Rancho Nuevo. Instead, by the mid-1980s more Kemp's ridley and other sea turtles than ever before washed up in the Texas Gulf Coast mutilated or dead—200 or 300 a year. This toll was devastating to species on the brink of extinction. Most of the dead turtles had suffocated in shrimp nets; others had died in collisions with boats or the boats' propellers. A substantial number of the dead turtles were females laden with eggs, multiplying the loss a hundredfold. In the midst of the carnage, in 1988, Allen and her student volunteers were upset to learn that the ten-year-old Operation Head Start had been canceled by the marine fisheries service and its parent agency, the U.S. Department of Commerce. Allen and her students had extended the life of the program several years more than it might otherwise have lasted; the power of publicity had held off the budget cutters for a time. But Head Start had finally become too vulnerable: None of the imprinted hatchlings had ever been observed to return to South Padre Island, despite regular patrols by Allen and her crew of HEART volunteers searching for signs of the marked turtles (or any Kemp's ridleys, for that matter). The experiment had been deemed a failure.

Allen pleaded for more time. "You need to be more patient. They will come back," she predicted. So little was known about the migratory patterns and wild behavior of these turtles that they might yet surprise us, she said. And even if they don't return, we are still giving an endangered species a chance for survival, and giving children an unparalleled learning opportunity. Her pleas were politely, but firmly, rejected.

And that, she recalls, led to a change in direction for HEART. Instead of closing up shop with the end of the program to bring life to the Kemp's ridley sea turtle, she instead focused on the industry that was bringing death to the species. A report by the National Research Council[2] had concluded that the 17,000-vessel shrimp trawling fleet in the Gulf of Mexico posed the single greatest threat to sea turtles, killing more of the animals than all other causes combined. Statistics compiled by the Department of Commerce showed that 14,000 turtles of all kinds were snared in shrimp nets each year in the gulf, and an estimated 4,000 of these were killed.

"There was really no choice," Allen would say later. "The commercial shrimp fleet was, despite their denials, driving the turtles into extinction. I knew it would be hard. But not as hard as watching a creature that has been on this earth for a hundred million years get wiped out in one generation."

Commercial shrimpers make their catch through bottom trawling, an indiscriminate form of fishing that wreaks havoc on ocean ecosystems, tearing up the seabed. Typical shrimp trawlers drag two forty-foot nets pulled from outriggers. The nets have boards attached to them that hold the mouths open and force the nets downward against the ocean floor, allowing them to scoop up bottom-feeding shrimp in huge numbers. But the nets gobble up everything else in their path— sharks, starfish, large game fish, turtles. Habitats and hatcheries are crushed in the process, and trawling is particularly damaging in the shallows, where it destroys whole sea life nurseries. Bottom trawling churns up sediment and pushes nutrients up in the water column, in turn promoting algal blooms. The algae are then consumed by hordes of bacteria, which strip the water of oxygen, creating dead zones where no fish can survive. In 2005, one such dead zone in the gulf reached the size of New Jersey.[3]

Just as the Endangered Species Act charged the Fish and Wildlife Service with protecting endangered land and freshwater species, the National Marine Fisheries Service had responsibility for protecting imperiled marine life. The agency prepared a recovery plan for the Kemp's ridley sea turtle that focused on the danger of being trapped in shrimp trawls. The plan championed a new device developed by the fisheries service, called a turtle excluder device (TED), which could be sewn into shrimp trawl nets, providing protection for the turtles without impairing the catch. The TED consists of thin metal bars in an oval frame that spans the net midway between the wide mouth, which scoops up the catch, and the other end: the "bag," where the catch accumulates. As shrimp flow in, they simply pass through the bars of the TED and into the bag. But larger and heavier creatures, such as turtles, strike the bars and they pop open to form a hatchway out of the net, ejecting the creature into the open water, then closing behind it. The ingenious device allows sea turtles to avoid being trapped and suffocating. (Kemp's ridleys don't actually drown; when they are submerged, involuntary muscles close their airway, so a turtle trapped too long underwater will suffocate.) A long series of tests by the fisheries service showed that a properly installed TED saved turtles 95 percent of the time, and allowed no more than 5 percent of a shrimp catch to escape. Because a TED also eliminated all sorts of by-catch—not just turtles but other unwanted creatures and debris—the shrimp catch came out of the water much "cleaner," and therefore could have about the same value as a slightly larger shrimp haul made without a TED.

Shrimp fishermen complained that the first version of the TED was too heavy, so the government developed a lighter model, which the shrimpers found too bulky. The government then offered a collapsible TED. The devices were distributed free to the fleet in a voluntary program beginning in 1982. But despite the modifications and pleas from environmentalists and government officials, only about 2 percent of shrimpers used the devices. The shrimpers hated the TED, and many fishermen swore they'd never use it, convinced that it would drive them to financial ruin and that the lost catch would be far larger than the government asserted; one industry study put the figure at 30 percent, six times the federal estimate. "There will be violence if they try to force this on us," one fisherman told a news reporter at the time. "We're on the edge, and that will put us over."

Allen and HEART were among the first to publicly assert that voluntary programs would not save the turtles. She began lobbying her congressmen, the congressional committees overseeing endangered species and fisheries, and the larger Washington-based environmental organizations with their big budgets and insider savvy, pushing for a federal rule making the TED mandatory. She spoke out at public meetings, wrote newspaper op-ed pieces, and testified at a series of rancorous federal hearings, where she was vilified by the shrimpers, who felt that their livelihood and their way of life were being threatened by this "turtle lady." Her argument that shrimping and species protection could coexist without driving anyone out of business simply did not jibe with the experience and traditions of the close-knit shrimping community.

The shrimpers rallied around Tee John Mialjevich, a blustery, 300-pound, six-foot-four advocate of the shrimp industry from Louisiana, who vowed neither he nor his supporters would ever obey a government edict that required them to use the TED. "We're not the culprits here," Mialjevich insisted, accusing Allen and other conservationists of a smear campaign. He said that in all his years of shrimping he had caught no more than six turtles, and that he had thrown them all back into the water alive.

Allen found those numbers doubtful at best, and certainly not typical—too many injured and dead turtles washed up on Texas beaches during the shrimping season, and then these numbers rapidly diminished once the trawlers departed. Allen also argued that "when a species is on the verge of going extinct, catching and possibly killing six of them is a disaster—particularly when there are close to twenty thousand other shrimpers out there in the gulf doing the same thing." She added, "Losing *one* turtle at this point is a disaster."

Allen had been a loyal, small-government Republican all her adult life, a believer in the wisdom of Ronald Reagan's view that the less regulation there was, the better—but these beautiful, endangered animals had given her new respect for the power of a well-placed regulation and the political will to enforce it. The voluntary approach had been tried. It had failed. Now, she argued, it was time to enforce the Endangered Species Act. For this, she was branded a traitor, a radical, a job-killer; she received hate mail, hang-up phone calls, and threatening calls. But she gave as good as she got, deriding fishermen who elevated a slightly higher profit margin for an unessential appetizer over the survival of an entire species. And when a letter to the editor questioned the purpose of saving turtles from extinction, Allen fired off a typically sharp reply: "In response to the question—*Exactly what is the purpose of a turtle?*—one might ask, what is the purpose of a bald eagle, a swan, a fish or a flower? Turtles don't have to justify their existence anymore than the writer of the letter does."

With Allen, HEART, and a coalition of national and state environmental organizations urging action, the National Marine Fisheries Service finally made the TED mandatory in 1988. Tee John Mialjevich and his supporters persuaded the state of Louisiana to sue the federal government on the shrimpers' behalf. ("Turtles don't vote," Governor Edwin Edwards explained.) The suit failed to remove the mandate, but the shrimpers persuaded Congress to postpone the effective date of the new regulations for another year. When the law requiring the TED finally took effect on July 1, 1989, Mialjevich led the shrimpers in a mass act of civil disobedience: more than 200 trawlers blockaded all shipping in the Houston waterways and blocked traffic into and out of America's largest oil depot for thirty-six hours. Tensions ran high, and the shrimpers threatened violence if they were forced to move; they even rammed a Coast Guard vessel that was shooting water cannons at the trawlers. And it worked. The administration of President George H. W. Bush (the first President Bush) began negotiations with Mialjevich, and Secretary of Commerce Robert Mosbacher decided to suspend the TED regulations and to consider alternatives—such as an unenforceable and largely useless requirement that trawlers lift their nets out of the water every 105 minutes to look for snared turtles. "Shrimpers Triumph Over TEDs," the next day's headline said. Mosbacher claimed that he had the legal right to suspend the rules—although no such power was granted under the Endangered

Species Act—because a "recalcitrant industry" had created a public safety emergency.

"I look to Secretary Mosbacher to enforce the law we worked for and not give in to mob action, which is what he's doing," a dispirited Allen told reporters.

Now it was her turn to sue, and HEART was joined by several larger environmental organizations as plaintiffs. A year later, they won reinstatement of the regulations. The shrimpers would have to use the TED, or face stiff fines, the feds promised. Within a year, the marine fisheries service was reporting that its new, tough attitude had produced better than 90 percent compliance. The major environmental groups who had come to Allen's assistance considered it a victory and moved on, leaving Allen to wonder why a combative industry that had been on the verge of insurrection would suddenly roll over. Hadn't the industry succeeded in backing down Mosbacher, the billionaire member of the first President Bush's inner circle, one of Houston's wealthiest and most influential citizens (who, in 2008, would become general chairman of Senator John McCain's campaign for president)?

"Something isn't right," Allen told her friends. "I saw the way they looked at me at those hearings. They don't give up that easily."

Then the deaths of sea turtles began to mount. The deaths should have declined with the TEDs in place, but instead, more and more of the turtles were washing up along the Texas coast. In 1994 there was a record high of 500, half of which were Kemp's ridleys. The stakes were especially high then because, in the early 1990s, the imprinted turtles that had been brought up from Mexico, hatched, and released from South Padre Island had finally begun to return to nest. First just a couple, then a dozen, then thirty-eight had appeared during the nesting seasons. Their Operation Head Start markings were unmistakable—and they were mixed in with other turtles that had not been imprinted.

Allen had been right when she had begged for a reprieve for the project—the long-lived turtles simply took until age thirteen to fifteen to begin nesting.

One of the junior biologists on the original Head Start project, Donna Shaver, now took charge, having made the Kemp's ridley sea turtle her lifework as a scientist. The new eggs would be protected from predators in a secure hatching area on the beaches, watched over around the clock (often by Shaver herself, sleeping on a cot on the beach). The hatchlings had to be released at the optimum time, when they emerged from their nest using all their energy reserves in a frenzy of swimming motions to propel themselves out to sea. Shaver found that hatchlings she had watched over in the old project were now coming back to give birth. She felt like a grandmother, she later told Allen.

At the same time, the population of Kemp's ridleys in Rancho Nuevo has surged to more than 3,000 nesting turtles each year, bolstered by the Mexican government's stringent protections of the eggs and the animals. But that progress, too, could be undone by the rising number of turtle deaths off the Texas coast.

Allen knew she needed outside help again. This time she contacted Todd Steiner—head of the Sea Turtle Restoration Project, based in San Francisco—who had worked many times with the Center for Biological Diversity on protecting sea life in the Pacific Ocean. Steiner was dubious at first, but soon realized from the data Allen provided that she was right: There were far too many turtle deaths, and the evidence suggested that these turtles had been killed by humans, not natural causes. Within a few months, HEART and Steiner filed suit under the Endangered Species Act, demanding that all shrimping in the gulf be halted until the enforcement of the TEDs regulations was improved. The government maintained it was enforcing the regulations and getting excellent compliance, but Steiner and Allen recruited the Humane Society to make an undercover investigation, which revealed that four out of ten ships in the shrimp fleet were not using TEDs. Large numbers of shrimpers were either removing the devices while at sea, or sewing them shut.

The much publicized investigation created yet another firestorm, and the victims, as usual, were the sea turtles. Mutilated turtle corpses began washing up on Texas beaches: flippers cut off, heads cut off, spikes driven through shells. One poor creature had been wrapped in chains and then thrown to its death, Mafia-style. The Sea Turtle Restoration Project and HEART offered a $5,000 reward for information about the culprits, and they began a public campaign for a permanent coastal preserve where shrimp trawling would be banned. HEART and other organizations also purchased billboards in Texas asking, "How many endangered Texas sea turtles get killed for your shrimp?"

In the end, the cruel tactics against the turtles generated a backlash against the shrimp industry among a public already unhappy about the earlier blockade. Hearings on creating the state-run coastal preserve that Allen had proposed began in 1998, and support for the idea appeared to be overwhelming—90 percent of public comments were favorable. Still, while the state continued to debate what to do, the turtle corpses kept washing up. There was no longer any question who was responsible: When a temporary ban on shrimp fishing was imposed for eight weeks in 1997, the numbers of dead and injured turtles turning up on beaches dropped dramatically, averaging not even two a week. Before and after the shrimping ban, strandings had been running between fifteen and twenty-five a week.

Allen and Steiner decided to attempt to shame the authorities into acting. They recruited protesters to wear turtle costumes and follow the governor—George W. Bush, who had launched his campaign for president in 1999. The protesters followed him throughout Texas and into California, relying on the assumption that as a candidate, Bush would want to do something to improve his abysmal record on the environment in Texas. Then HEART and the Sea Turtle Restoration Project bought two full-page ads in *The New York Times*, decrying the plight of the sea turtle and Bush's inaction. One of the ads had the heading, "If Governor Bush doesn't save the Texas sea turtle maybe President Gore will." Two days later, Bush deputized seventy game wardens to assist with patrolling the gulf and enforcing the federal shrimping regulations. The feds, meanwhile, finally began to enforce the law vigorously, for the first time charging a shrimper with a criminal violation of the Endangered Species Act instead of doing the equivalent of writing a speeding ticket. In a criminal case, the trawler, the equipment, and the catch are seized. Instead of a fine, now one violation of the TED regulations could put a shrimper out of business. In a few months, compliance with the TED requirement increased from 60 percent to 75 percent.

By 2000, officials in Texas had agreed to a partial coastal preserve: Shrimping would be banned within five miles of the coast between December 1 and July 15, and there would be no night shrimping at all. There were also other restrictions on trawling, intended to complement federal protections for sea life. It was not the year-round ban in shallow coast waters that Allen sought and that 96 percent of public commenters at the hearings had endorsed, but it was enough to keep the trawlers away during the nesting season.

The shrimpers remained bitter, but accepted the regulations at last, and they have gotten some relief through legislation that limited the import of foreign shrimp. By law, shrimp can be imported only from nations that also require TEDs.

After working to protect the turtles for more than twenty-five years, Allen still remains vigilant, organizing teams of volunteers to comb the beaches during nesting season, searching for the turtles' distinctive half-moon tracks in the sand. In 2007, 128 Kemp's ridley sea turtle nests were counted—a record—and protected through Donna Shaver's program and by Allen's volunteers. HEART is no longer Allen's little nonprofit; it's now the Gulf Coast branch of the much larger Sea Turtle Restoration Project, and Allen's work there is now her job, after many years of volunteerism. She will not give her age, saying only that she is a "pre-boomer," but her enthusiasm for turtles, for being present during the hatchings and releases, is undiminished as she continues to work for more rigorous protections for the Kemp's ridley, the species she saved from extinction.

"Turtles don't do anything quickly," she says. "So when you are into turtles, you learn to be patient, too."

14.

wild man

Robert Edward Turner III is a different sort of eco baron: wealthy, privileged, controversial, famous, and infamous. His philanthropy sets a very high bar for other Americans with big bank accounts, as he has fashioned himself into the nation's most generous and flamboyant donor to public causes. Yet he is known to the American public far less for his groundbreaking wildlands preservation, restoration, and re-wilding than for making provocative and occasionally outrageous public statements. It wasn't his large private conservation projects—which dwarf whole countries' efforts—that generated headlines in 2008. It was his off-the-cuff remark on a public television show that created a firestorm when he predicted that continued inaction against global warming would doom humanity to social decay, starvation, and a great many deaths in thirty or forty years. Then, in a deliberately over-the-top aside, he suggested, "The rest of us will be cannibals."

His many detractors pounced, and they included conservative

activists, one-world-government conspiracy theorists, haters of Jane Fonda, and deniers of global warming. Turner's gift for public hyperbole puts him in the limelight, but it also tempts his critics to write him off as a liberal lunatic: There's a reason, after all, that Ted Turner is nicknamed the "mouth of the South."

But there's also a reason that Doug Tompkins thinks Turner could go down in history as "his generation's giant of wildlands philanthropy," and that Mike Phillips, who was once the National Park Service's expert on wolves and now works for Turner, says that the founder of CNN, the twenty-four-hour cable news network, has surpassed Rockefeller "with a body of conservation work that exceeds any human who has ever lived."

Ted Turner owns more land in America than any other single individual—about 2 million acres, most of it made up of fifteen immense ranches sprawled across six states, primarily in the West. When the federal government needs help reintroducing nearly extinct species to their old ranges—wolves, ferrets, bears, trout, woodpeckers—it turns to Turner. His mission is as simple as it is unprecedented: As much as practically possible, he is attempting to return his lands to the natural state in which they once existed, before man came along.

"We're trying to replace as many missing pieces to the environment as we can," Turner says. "We're trying to save what we can of the natural world."

Ted Turner, who turned seventy in November 2008 and had a net worth of $2.3 billion, has a mind-boggling résumé as businessman, media mogul, outdoorsman, environmentalist, and philanthropist, not least because he lost most of his fortune—"Six to eight billion, depending on what time of day it is," he recently quipped—in a disastrous corporate merger with tech giant turned dinosaur, America Online (AOL). It was one of very few big missteps in a remarkable career, and though it slowed the pace of his philanthropy (his

Turner Foundation had to retreat from $45 million in grants for environmental groups in 2000 to $8.6 million for a variety of environmental and social causes in 2007[1]), he says he still spends more on his wildlands conservation and philanthropic grants than his annual business income brings in, routinely maxing out his tax deductions for donations. He figures he's given away $1.5 billion since the mid-1990s. In 1998, he pledged an unprecedented $1 billion donation to the United Nations to support a broad mix of causes, including his two highest priorities: nuclear disarmament and the battle against climate change. The donation was made in installments, with the billionth dollar delivered by Turner's nonprofit United Nations Foundation in 2006.

Turner is a typical self-made man. He was a rebellious child—a fact that surprises no one who has known him as an adult—and he was shipped off to various boarding and military schools. The discipline imposed on him there (one school lost count of his demerits) did little to suppress his defiance of authority and conventional wisdom. He was expelled from Brown University in 1960 for sneaking a woman into his dormitory, and he came home to work for his father's billboard company, Turner Outdoor Advertising, in Savannah, Georgia. When his father committed suicide in 1963, Turner, then twenty-four, took over the business, which was worth about $1 million, though it was heavily in debt. Turner discovered that he liked being in charge, threw himself into the work, and built a media empire. At the time, he summed up his philosophy in the least altruistic terms imaginable: "Life is a game. Money is how we keep score."

In 1970, Turner decided to expand from billboards into television by buying two properties: a small, money-losing UHF television station, and the broadcast rights to a library of vintage black-and-white movies and old television shows. At that time, television was dominated by the three broadcast networks and their local affiliate stations, and the recently added UHF stations were obscure and hard for viewers to find. But Turner became one of the first to parlay a small station into a national cable operation, offering old movies and cartoons to an audience hungry for alternative programs. Turner was one of the first independent broadcasters to use satellite technology, and then the explosive growth of the cable industry, to reach a huge viewership. In the mid-1970s he used his cable profits to buy the Atlanta Braves baseball team and the Atlanta Hawks basketball team and added their games to his national broadcasts, transforming his operation into the nation's first "superstation." The Turner Broadcasting System—TBS—was born.

In 1980, he launched Cable News Network, which industry insiders predicted would be a disaster. No one wanted twenty-four-hour news, it was said. Cocky, blustery, and profane, Turner told the so-called experts to go to hell. CNN made Turner a billionaire and permanently ended the predominance of the network nightly news. He then purchased MGM/United Artists Entertainment and baffled the entertainment industry by selling the Hollywood studio back to its original owner. He's crazy, the experts said again. Go to hell, Turner answered again. For he had kept one piece of the operation— the studio's enormous library of films, including such classics as *The Wizard of Oz, Citizen Kane,* and *Gone with the Wind* (which came with the actual Academy Award statue the film had won). Turner had anticipated the value those films would have in the new but rapidly expanding market for home videos. And his new Cartoon Network cable station now had a huge library of classic cartoons, making it a destination for children around the country. He made $120 million from that film library in the first year. In 1995, he merged his cable, broadcast, film, and sports holdings with the media giant Time Warner, becoming its largest individual stockholder, a vice chairman, and a member of the board of directors—and multiplying his wealth five times over. Most of that money vanished after he went along with the merger that combined Time Warner, the "old media" giant, with "new media" AOL, the former king of dial-up online access, in 2001. The ink had barely dried on the merger when it became clear that it was a terrible mistake: AOL was dying, losing subscribers from its online "walled garden," being made irrelevant by the advent of easy access to the World Wide Web and the rise of broadband Internet service. Stock in the company, briefly dubbed AOL-Time Warner, tanked; "AOL" has since been erased from the company name, and Turner left the company.

Along the way, Turner married three times, most recently the actress and activist Jane Fonda (they were married in 1991, divorced in 2001); had five children, all of whom are involved with their own environmental activism, as well as sitting on the board of the Turner Foundation; skippered the winning yacht in the America's Cup competition in 1977; gave the Atlanta Braves' manager the day off and managed the team himself for a game; was barred from the Braves' dugout by the commissioner of baseball; founded the Goodwill Games; and outraged conservatives and churchgoers by calling Christianity a "religion for losers"—a remark for which he later apologized. He then joined a $200 million partnership with Lutheran and Methodist church groups to create an antimalaria program in Africa. What he has seldom revealed was that he lost his faith at age twenty, when his teenage sister died of lupus after a long illness; Turner had prayed for her daily for years and felt he had been a dupe. His daughter, Laura Seydel, once said he felt driven to save the world because he had decided God wouldn't do it, though in recent years he has said he has begun to pray again. "It doesn't hurt," he allows.

Over time, his philosophy that "life is a game" seemed to shift, and he began to speak of doing something meaningful with his wealth. He claimed a new motto: "I don't give till it hurts. I give till it feels good." Turner had for many years defined himself as "a screaming ecologist and a wild-eyed do-gooder," and he stepped up his efforts in the 1990s, when he began buying land in the West and decided that restoring it to its natural state was his calling. Much of his ranch land has been placed in conservation easements that ban most development.

In 1997, he created the Turner Endangered Species Fund and hired Mike Phillips away from the U.S. Fish and Wildlife Service to be the new project's first director. Phillips, a biologist from Montana, was the first project manager of the successful program to release gray wolves into Yellowstone National Park, at a time when the wolf was virtually extinct in the wild. Turner's youngest son, Beau, is the fund's president and manages the wildlife programs. The Turner Endangered Species fund specializes in reintroducing endangered species, including wolves, bears, and other predators, to their original habitats. The fund also focuses on the preservation of two vital ecosystems: grasslands and pollinator corridors (habitats for migratory doves, bats, hummingbirds, and butterflies) that are collapsing from development and pollution, though they are essential parts of the food chain. And at many of his ranches, projects for eradicating nonnative plants species and restoring native ecosystems are also in progress; they include the planting of thousands of native tree seedlings.

Turner's Flying D Ranch, 113,000 acres in southwestern Montana, is the largest private landholding in the Yellowstone Park ecosystem, where wolves, elk herds, and grizzly bears safely roam and hunt, moving unimpeded between his property and the adjacent sprawling public forests and parkland. Turner had all the old ranch fences pulled down to allow wildlife to move freely, and he incorporated his ranch into the Yellowstone-to-Yukon Conservation Initiative, an affiliate of the Wildlands Project. The initiative seeks to stitch together 1,800 miles of public and private land into a network of wildlife corridors and reserves from Big Sky country to the Canadian tundra—an enormous international zone of sustainable, unsullied nature intended to resemble what the region looked like before the West was settled.

Turner has been something of an absolutist on this point of restoration: In partnership with the state of Montana, he is reintroducing the nearly vanished westslope cutthroat trout to its native Cherry Creek, which runs through his ranch and onto adjacent public lands.

The westslope is Montana's state fish, but its numbers have dropped drastically because nonnative rainbow and brown trout have been placed in Montana waters by fishermen. The hardier nonnative species crowded out the Montana fish. Because the Flying D's fifty-mile portion of Cherry Creek is blocked at one end by a waterfall, state biologists realized it would be possible to isolate its length from other portions of the creek with nonnative fish, making it a perfect laboratory for restoration. Turner, an avid fly fisherman, agreed to the project, even though it meant the nonnative fish on his property had to be killed with a quick-acting, specifically targeted, fast-dissipating poison before the westslope could be placed back in Cherry Creek. This poisoning caused an understandable uproar among locals and led to five years of lawsuits, but Turner, supported by the state of Montana, was adamant: he would put nature back the way it was supposed to be.

In New Mexico, Turner's 250,000-acre Ladder Ranch is a similar laboratory of competing interests. First he reintroduced the imperiled desert bighorn sheep to his lands. Then, after the sheep were well-established, wolf packs were re-wilded there, too, even though some of the very expensively re-wilded sheep would become wolf food. But that's what real nature is about—predator and prey, completing the cycle, according to Turner.

The Ladder Ranch is a major connector in another arm of the Wildlands Project, the Sky Islands Wildlands Network, which straddles Arizona, New Mexico, and northern Mexico. Sky Islands is an attempt to link wildlife corridors, reserves, and "islands" of mountainous habitats spanning four diverse ecosystems: the northern temperate Rocky Mountains region, the southern subtropical Sierra Madre, the Sonoran Desert, and the Chihuahuan Desert. The area is a major repository of biodiversity, including half the bird species in North America. Ladder Ranch also has a captive breeding facility for Mexican wolves destined for release on Turner's land and federal lands. Some of Turner's wolves have been reintroduced at Kieran Suckling's and Peter Galvin's old stomping grounds in the Gila National Forest, amid protests and threats from locals. The re-wilding of wolves, which Turner considers an essential part of restoring a healthy land-scape, is very controversial among ranchers, who fear their livestock will be killed and their livelihoods threatened. Turner has called for cooperation to preserve an important endangered species, but it has been hard to come by, and some reintroduced wolves have been shot. In 2007, an entire reintroduced wolf pack vanished from the Gila. Turner, meanwhile, has expressed the hope that he will hear the howl of a completely wild wolf at the Flying D before he dies.

At the Vermijo Park Ranch along the New Mexico–Colorado border—Turner's largest property holding, at 580,000 acres—prairie dogs rule. Colonies of these highly social burrowing rodents—typically exterminated as pests by ranchers—are allowed to thrive and dig everywhere at Vermijo. They are not officially designated as endangered, although prairie dogs now live on only 2 percent of their once enormous range across America's grasslands, which themselves are imperiled. Turner lets the prairie dogs flourish at the ranch be-cause of their role as the primary prey for an animal that is nearly extinct in the wild: the black-footed ferret. The ferrets are raised wild on the ranch and used in reintroduction programs throughout the West. The wild-raised ferrets do much better once released than fer-rets raised in zoos. The prairie dogs and ferrets are part of a larger effort to restore prairie landscapes; biologists working for the Turner Endangered Species Fund are also studying the possible reintroduc-tion on the ranch of imperiled falcons, condors, spotted owls, and the Rio Grand cutthroat trout.

The ranch houses another of Turner's captive wolf breeding fa-cilities, this one for the Southern Rocky Mountain wolf, a species being re-wilded slowly in the Rockies. Ranchers in New Mexico have vigorously opposed the reintroduction, citing livestock deaths in the Yellowstone area after wolves were reintroduced there, and making Turner something of a pariah among some of his ranching neighbors.

Turner didn't help his case by saying in 1997 that raising cattle in arid environments such as New Mexico was foolish and unsustainable. "He has thumbed his nose at the custom, culture, and ethnic diversity of the state," the head of the New Mexico Cattle Growers said in response.

But Turner has a point: The cattle industry is responsible for 18 percent of all greenhouse gas emissions worldwide (not so much from the animals themselves, but from related deforestation to expand pastures and to grow feed crops, as well as the energy-intensive process of producing chemical fertilizers used for that feed).[2] In the last five months of 2007 alone, more than 1,200 square miles of Brazil's rain forest were burned and clear-cut to make room for feed crops and pasture; Brazil's president imposed a state of emergency to halt the destruction. Beef, like oil, is subsidized by the U.S. government, which makes it economically viable to raise cattle in arid climates such as New Mexico, and enables America to eat the most beef in the world: 5 percent of the population consuming 15 percent of the meat.[3] If Americans would simply reduce their meat consumption by one-fifth, according to an analysis by two geophysicists—Gidon Eshel of the Bard Center and Pamela A. Martin of the University of Chicago—this would have the same effect as if all Americans traded in their cars for superefficient, low-polluting Priuses.[4]

Although he hasn't changed his mind about the environmental problems associated with the beef industry, Turner has stopped publicly criticizing cattlemen, instead offering a personal counterexample. He has removed cattle from the ranches he buys and instead raises bison there—more than 50,000 head, the largest herd in America— and uses their meat at his chain of restaurants, Ted's Montana Grill. Buffalo are indigenous to the plains and prairies of the West, Turner says, and do not inflict nearly as much damage on the ecosystems as cattle herds do. Through his restaurant chain, which had grown from one restaurant in 2002 to more than fifty in eighteen states by 2008, he has been championing some modest green business methods—alternative energy, conservation, even talking a manufacturer into once again making straws out of paper instead of petroleum-based plastic. He says he hopes to demonstrate that operating ranches and businesses in an environmentally sound fashion, including nurturing endangered species on private lands, can be a profitable enterprise. He allows limited hunting and fishing on his properties, for a fee that starts at $400 a day per guest, and that can top $12,000 for a weeklong hunting excursion on the breathtaking Vermejo Ranch.

Apart from conservation efforts on his ranches, Turner has dispensed as much as $60 million a year since 1990 to environmental groups in the United States. Like Doug Tompkins's Foundation for Deep Ecology, he has often sought effective, upstart organizations to support financially, as well as larger, more established groups. For many years, the former director of Greenpeace USA, Peter Bahouth, administered the Turner Foundation grants program, though he has since moved on to direct the U.S. Climate Action Network. At one point, Turner was supporting nearly 500 different environmental organizations—among them the Center for Biological Diversity at a time when it was struggling to keep afloat. Turner has doled out grants to groups and programs protecting rivers, otters, wolves, bears, dolphins, rain forests, ducks, sea turtles, whales, pine trees, manatees, monarch butterflies, jaguars, grasslands, cactus forests, and salmon—to name just a few—along with groups promoting renewable energy, environmental justice, bicycle riding, the Boy Scouts, and an end to mountaintop removal methods of coal mining. The description of programs seeking funding from Turner has filled as many as three bound volumes each year.

Bahouth would spend part of the year researching and visiting environmental groups that had applied for grants, then assembling those volumes, which were, in effect, an analysis of the state of American environmentalism, along with the pros and cons of supporting each applicant. The volumes would be distributed to Turner, his five adult children, and (while she and Turner were married) Jane Fonda, so they could vote on which projects to support. Each board member also had a discretionary fund for his or her own pet projects; these included, for example, sustainable home building groups and a local "riverkeeper" group.

Typical of the organizations Bahouth discovered and the Turner Foundation supported was Ozone Action, an obscure outfit advocating causes related to climate change. It was run by John Passacantando, who later took over Bahouth's former position as director of Greenpeace USA. The Turner Foundation gave Ozone Action $170,000, in part to fund its hounding of presidential candidates on the campaign trail in 1999 and 2000. Ozone Action representatives would show up at campaign events wearing big hats shaped like smokestacks and carrying signs asking, "What's Your Plan?" The goal was to jump-start candidates into addressing climate change, a topic that was being avoided like the plague during the campaign, even with Al Gore leading the Democratic primary races. Sometimes the Ozone Action people were ignored; at other times the Secret Service ejected them. But after seeing them at half a dozen events, Senator John McCain tired of being harangued and invited them on board his campaign bus. The conservative Arizona Republican listened to their pitch about the need for urgent action on climate change, asked his staff to check out what they had told him. Pretty soon, a senator who hadn't spent much time or political capital on the issue of global warming was holding hearings, sponsoring legislation, inviting Ozone Action's director to testify, and, for his presidential run in 2008, developing a program for dealing with climate change that many environmental organizations considered flawed, but vastly better than those of his competitors in the Republican primaries.

"So now people say, hey, McCain is pretty good on global warming," Bahouth chortles. "Why is he good? Because we funded a group that bird-dogged him on the campaign trail. What's that investment worth?"

The billions that Ted Turner lost because of the disastrous merger with AOL forced him to cut back his grants program, and to transform the once wide-open application process into an invitation-only approach. When he broke the news to his children, the fund's trustees, he wept.

He says he has no plans for retirement in his seventies. He has hinted that he will donate most of his vast real estate holdings to become "national parks or something" after his death; whatever form this donation takes, he has vowed that the bulk of those 2 million acres—more, if he can swing it—will be permanently preserved as wilderness. Between his wildlands conservation project, his work for endangered species, and his eco-philanthropy, Turner is arguably America's most prolific environmentalist. He has received accolades, awards, and honorary degrees, and has often appeared on magazine covers, including *Time* Magazine's Man of the Year cover in 1991; but at times he seems to be portrayed and perceived more as robber baron than eco baron, viewed with suspicion and partisan distrust.

The "cannibal controversy" was typical. It arose from a couple of sentences during an hour-long television interview in April 2008 with the PBS host Charlie Rose. In that interview, Turner discussed his views on the most critical world problems, which he ranked in importance: proliferation of nuclear weapons, global warming, overpopulation, ocean and air pollution. In his typically colorful, hyperbolic manner, Turner portrayed these problems as a grave threat that could bring down civilization. But he also argued that resolute action to solve these problems would create an unprecedented opportunity for jobs, wealth, and security:

"Fossil fuel's day is over. We're poisoning our children and ourselves. . . . We have to mobilize the same way we did when we entered World War II in 1941. We have to fully mobilize everything we have in changing the energy system over. It is going to be the biggest business project in the history of the world. Fortunes, billions of dollars, are going to be made. Hundreds of thousands of people are going to be employed. We're going to have clean air. We're going to have so many benefits, it's not going to cost us anything once we get going. . . .

"Not doing it is going to be catastrophic. We'll be eight degrees hotter in . . . thirty or forty years and basically none of the crops will grow. Most of the people will have died and the rest of us will be cannibals. Civilization will have broken down. The few people left will be living in a failed state—like Somalia or Sudan—and living conditions will be intolerable. The droughts will be so bad there'll be no more corn grown. Not doing it is suicide. Just like dropping bombs on each other, nuclear weapons, is suicide."

The next day, over and over throughout the media and the Internet, the snippet about "cannibals," which had lasted only twenty seconds, was replayed, the focus of coverage of Turner's remarks. He was talking "wild-eyed lunacy," one columnist wrote. Another called him "crazy Uncle Ted." He was said to be spreading "neo-atheist-Marxist ideas." Within three days, a Google search for Turner's name coupled with the word "cannibal" produced 10,400 hits.

Aside from removing the remark from its context, which would have made it clear that Turner was trying to paint a worst-case scenario that he did not believe would ever occur, the commenters failed to realize that Turner was alluding to a recent best-selling book, *Collapse*, by the evolutionary biologist and Pulitzer Prize–winning author Jared Diamond. The book analyzes the causes behind the fall of past civilizations, then contrasts those factors with the challenges facing humanity today. Easter Island was a particularly vivid example cited by Diamond: a once vibrant society that exhausted its resources, devolved into warring factions, and finally resorted to cannibalism as the only alternative to starvation. Human history offers many such examples of resource depletion followed by the fall of cultures and descent into cannibalism—it's ugly, it's unpleasant, no one likes to think about it, but it happens to be true. Turner was simply suggesting that history could repeat itself.

None of his critics and none of the news reports on the interview included what Turner had said next to Charlie Rose, when he made it clear that he did not believe his apocalyptic scenario would ever come to pass, expressing a hopeful outlook to explain why he felt compelled to expend so much of his wealth on environmental causes:

"We're in a tough situation, but we can play our way out of it. . . . I love this planet and it's worth saving. I know we're the same people who did the Holocaust, but we're also the people who did the Mona Lisa and Beethoven's Fifth Symphony. This world, we can't turn it into a cinder, for ourselves and for our children. And it's worth fighting for."

schemers and dreamers

We could have saved the earth, but we were too damned cheap.

—KURT VONNEGUT JR.

For many years, the National Aeronautics and Space Administration (NASA) sponsored a contest called SpaceSet, in which high school students designed elaborate colonies for thousands of men and women on Mars or the moon, using real science and existing technologies. These annual competitions produced some remarkable examples of innovative engineering, ecology, industry, and sociology, as student teams worldwide spent months collaborating on and modeling these cities in space.

The winning entries all tended to share some basic characteristics worth noting.

These extraterrestrial colonies did not equip their citizens with vehicles that filled the air supply with toxic and cancer-causing fumes.

They did not rely on inefficient transportation technology that wasted 80 percent of the energy it expended.

They did not rely on finite supplies of fuels that had to be extracted from the ground thousands of miles away from the colony, then transported at great expense aboard massive, leak-prone tankers.

They did not place colonists' homes and workplaces many miles apart, requiring long commutes and expenditure of energy.

They did not permit the use of archaic and wasteful lightbulbs and appliances based on obsolete technology.

They did not offer tax benefits to encourage wasteful and polluting transportation technologies over efficient and clean ones.

They did not dump their chemical and biological wastes in the colony's water, food, and air supplies.

In other words, the most successful space colony designers looked at how human civilization works on earth, and pretty much did the opposite.

Sometimes in order to comprehend a problem, a little distance is needed (in this case, the 238,400 miles separating the earth from the moon): The way we live would quickly be fatal in the self-contained environment of a space colony. Clean, efficient technology, zero emissions, and recycling of waste isn't just a good idea when settling the moon or Mars, but a matter of life and death.

This has turned out to be no less true on earth, notwithstanding long years of denial and a naive misconception that the world offers limitless bounty. As the engineer, philosopher, and visionary Buckminster Fuller noted in 1963, in describing the planet as "Spaceship Earth," the world's resources are just as finite, and just as subject to destruction, exhaustion, and contamination through man's carelessness, as any space colony we might imagine or build.[1] The terrestrial scale may be larger, making the effect slower, more diffuse, and easier to shrug off or blame on something else, but the principles are the same. The value of the metaphor of "Spaceship Earth" is its ability to make clear what we normally miss: that if it would be insane to live on Mars as we live on earth, then it must also be insane to live as we do on earth itself.

This fundamental notion—that we have constructed an unsustainably wasteful way of living that makes sense only because it is all we know— is what the eco barons understood early on. As Doug Tompkins's river guide turned Wal-Mart consultant, Jib Ellison, puts it: "It's like we're denying gravity. We've built all these machines and expended all this energy, just to live a foot off the ground. And we keep doing it, year after year, until someone finally says, Why?"

This is why Doug Tompkins abandoned his fashion empire, found a rugged cabin in the middle of Eden, and started saving and restoring paradise, one plot, one fence, and one tree at a time. This is why two owl hooters lived like monks and found a way to use the law to save forests, species, and clean air when no one else had the temerity to tell government and industry: *Enough!* It is why a professor and his students build cars that burn no gas, why a cosmetics queen is spending her fortune to save the last great forest of our forefathers, why a media mogul lies awake at his ranch and listens for the return of the wolves, why a "turtle lady" walks along a beach each spring and waits, her heart in her throat, for that homely, beautiful, beaked face to appear out of the waves and for the mother turtle to lay her precious eggs in the wet, warm sand. They see, clearly, that what we're doing as a society is not working. Their response is not to shout about it, or lobby about it, or generate self-aggrandizing headlines about it. Their response is to do something about it, and their results have been spectacular.

There are thousands of environmentalists and activists doing important work in America and around the world. But a few of them go farther—these dreamers, schemers, moguls, and coupon clippers; these eco barons. There are others out there, certainly, more all the time; this book is by no means an exhaustive list, but it is an inspiring selection. The eco barons depicted here stand out because they are game-changers, accomplishing something extraordinary, raising the bar of the possible, usually after being told that what they are attempting is impossible.

They have undertaken an epic project: to save a piece of Eden, to restore it and nurture it, and thereby set an example for the rest of us. Their actions are their message: that there is a clear choice, a difficult choice, a right choice, and to make it is to express the faith that it is not too late to save the world—and that a new way of living can be better, healthier, smarter, and more prosperous. They argue that it should never be too late for the America that helped save the world during World War II and then rebuilt it afterward—for friend and foe alike. How can it be too late for the America that promised the moon in ten years—and got there in nine? Or for the America that invented the lightbulb, the laser, the personal computer, and, most gloriously, our Constitution? The eco barons say it is not too late—if we heed and follow their example in the present, and the example of triumphs from our past.

The problem is time. There was time to act in 1980, but then Ronald Reagan blocked the road to energy independence and squandered America's lead in solar and wind energy, because that is what the special interests of oil and autos demanded. There was less time, but still there was time, in 1997, when Bill Clinton and Al Gore signed the Kyoto Protocol. But then the plug was pulled by a craven Congress, because that is what the special interests demanded. There was less time, but still there was time, when George W. Bush took office in 2001, but instead of acting in America's best interests, he allied himself once again with a few special interests—big oil, big auto, big coal. He dismissed, vilified, and censored science and squandered eight years by fighting every attempt to combat global warming that came along.

There is now no time left. The train has left the station. The climate is warming. The best we can do now is minimize the impact. "The only resolution is for humans to move to a fundamentally different energy pathway within a decade," according to James Hansen, director of NASA's Goddard Institute for Space Studies and a leading climate scientist—one of those Bush attempted to censor. "Otherwise, it will be too late for one-third of the world's animal and plant species and millions of the most vulnerable members of our own species."

It took hundreds of millions of years for nature to construct the storehouses of coal, oil, and gas beneath the surface of the earth, the chemical residue left behind by countless extinct prehistoric life-forms (which is why it is called *fossil* fuel). Oil and natural gas formed from deep pressure and heat exerted on huge layers of dead plankton and algae buried beneath the bottoms of ancient seas, some of which are now the world's great deserts. Coal is made of the fossil remains of primordial forests buried by time and cataclysm. Every time a gas tank is filled or an electric generator is fired up, the remains of long-extinct creatures are providing the energy.

What nature was doing for all those millions of years was seques-tering carbon, which is now being released by humanity in fantasti-cally huge amounts: 75 million metric tons a day. Hydrocarbons once safely buried in the ground are becoming carbon dioxide not so safely altering the air and climate. In the case of oil and gas, we are well on the way to reversing, in little more than a century, what nature took 500 million years to accomplish. The data are undeniable: Concentra-tion in the atmosphere of the principal greenhouse gas causing global warming, carbon dioxide, was below 290 parts per million for 1,000 years leading up to the industrial revolution in 1850, as shown by the indelible record preserved within polar ice core samples. By 1900, the concentration exceeded 300 parts per million. In 2008, it reached 385 parts per million.[2]

The threshold at which the weather starts to change into some-thing far less hospitable to the plants and animals we depend on is thought to be 350 parts per million. That threshold has not been crossed significantly in the past 400,000 years—until now. We are living in the red zone. Life can evolve and adapt to gradual changes in climate. But the last time climate shifted as rapidly as man is now causing it to warm was about 65 million years ago, when a massive meteor hit the earth, and the dinosaurs went extinct.

The cause is different this time, but "Spaceship Earth" is again being poisoned so fast that its systems cannot purify and recycle the environment on their own. Pollution, toxins, deforestation, habitat loss, and record levels of oil and coal use are creating dire conditions that, absent immediate change, will be very hard to reverse. Animals are already going extinct at high rates. Weather is growing more extreme. Do we know that the devastation of Hurricane Katrina in 2005, devastation from which the Gulf Coast had still not recovered four years later, was caused by climate change? No, this cannot be said with certainty, because proving causation for specific weather events is beyond our technical prowess. Do we know that Katrina was part of a larger pattern of extreme weather that is mostly driven by climate change? Yes. The scientific community is in broad agreement about this. There is considerable debate over details—global weather being one of the most complex systems to model, even with supercomputers—but with regard to the big picture, there is little doubt among climate scientists that the planet is in trouble and that we desperately need immediate strong leadership, regulatory reform, and a vigorous campaign to change the way we obtain and use energy.[3]

In 1990 a renowned Swedish cancer researcher, Dr. Karl-Henrik Robèrt, joined by fifty other scientists, developed a four-step program outlining how a country could be transformed to embrace sustainability in every sector—business, government, industry, and the daily lives of ordinary citizens. The goal would be to reduce pollution, overharvesting, overfishing, exposure to toxins, and greenhouse gas emissions, while eliminating dependence on fossil fuels and shifting to renewable energy. The plan's most basic idea was simple and profound: Society should take from the earth only those resources that can be renewed, and emit only those substances that can be safely absorbed or purged by nature. Robèrt called the plan "The Natural Step."[4]

In America, such seemingly utopian ideas tend to sink like rocks. But what if, against all expectations, a president promoted and embraced this plan as a new national project, launching it as President Kennedy launched the race to the moon? What if a readable, digestible version of the report and "The Natural Step" was mailed to every home and school in the country, and national discussions and classroom lessons on sustainability ensued? Imagine celebrities, artists, and broadcasters promoting this new idea on television and radio, urging the country to incorporate these principles into everyday life and work as vital to the nation's security and prosperity. What would happen?

In Sweden's case, all of this did happen (except that it was the king, not a president, who embraced the idea). "The Natural Step" was launched nationwide in Sweden in 1990 as goals the nation should aspire to achieve, through rigorous emissions controls on cars and through the use of alternative energy. The results have been spectacular: Sweden has become the most sustainable nation on the planet. It is lowest in a widely followed annual international survey conducted in Germany of the fifty-three largest emitters of greenhouse gases.[5] The United States ranks fifty-second on that list, with only Saudi Arabia deemed a worse contributor to climate change. Sweden has reduced its greenhouse gas emissions by 9 percent below 1990 levels, an achievement that so far exceeds the mandates of the Kyoto treaty, whereas the rest of the world has continued to increase emissions. Sweden enacted a plan to remove oil completely from its economy by 2020, and so far has achieved a 70 percent reduction in the use of oil for home heating and has held industrial consumption at 1994 levels. The major challenge remains transportation; but already, almost one-third of Sweden's energy comes from renewable sources.

In the United States, the argument against taking such drastic action to combat global warming and wean the country from oil has always been that it would cause severe economic hardship and put

American industry at a competitive disadvantage with the rest of the world.

Sweden's economy, home to Ikea, Saab, Electrolux, Ericsson, and Volvo, expanded 47 percent between 1990 and 2008.

In the developed world, disagreement with Sweden's notion that society must be transformed to meet the challenge of climate change and the destruction of nature emanates most strongly from America. Craven leaders, partisan politics, and junk science have fueled and confused a debate in which the fundamentals, if not the details, should have been settled long ago. The result is a public that doesn't really get global warming. People have heard of it, they know what the term means—Al Gore and his book and film *An Inconvenient Truth* deserve credit for that. And yet, polling by the Gallup organization shows that only one-third of Americans in 2008 believed any urgent national action was necessary to deal with climate change. Similar polls suggest that only half of Americans even accept the idea that humans could be causing climate change. More Americans believed the falsehood that Iraq was behind the attacks of 9/11 than accepted global warming as an imminent threat. More than anything else, this shows the power of the president: A large majority of Americans believed the lie about Iraq's being a threat to U.S. security because the president—any president—has an immense ability to sway public opinion when he so chooses. In the aftermath of 9/11, George Bush could have used that power to mobilize the country to end our oil addiction, to launch an urgent program for renewable energy, and to muster the nation's resources against climate change. He might then have gone down in history as the greatest eco baron of all time, as the president who saved the earth. But he did none of those things, and such inaction from a president has power, too. It is the reason America lags far behind the rest of the industrialized world in the battle against global warming and why, if Bush is remembered in future decades as he is perceived today by many historians and members of the public—as the worst president in the nation's history—it may well be his failure as an environmentalist that is regretted most by the future generations who must live with the consequences.

Meanwhile, the work of the eco barons continues.

The Center for Biological Diversity continued to shift its focus, putting global warming and its disastrous effect on endangered species near the forefront of its efforts. The center created a new climate institute, run by Kassie Siegel, who attained national prominence through her championing of the polar bear to force a national response to global warming. That breakthrough accomplishment was threatened by a frenzy of last-minute "midnight rule-making" by President Bush, who in his last weeks in office hammered through a series of regulations that sought to gut the Endangered Species Act. The new rules would remove protections from the polar bear and end consideration of global warming as a cause of extinction in all cases—while accelerating coal mining and oil drilling in protected oceans and wilderness areas. The center, along with Greenpeace and Defenders of Wildlife, immediately sued to block the new regulations on the grounds that they violated the endangered species law they were supposed to implement, although the legal thicket erected by the administration lawyers seemed likely to cause a costly and protracted court battle. "They're trying to accomplish by fiat what they couldn't do with the courts or Congress," said Karen Suckling. "It can't be allowed to stand."

In Maine, Roxanne Quimby bought more land for preservation. The huge infusion of cash from the sale of her remaining shares of Burt's Bees to Clorox provided everything she needed to realize her vision of a permanent preserve or park in the Maine Woods, although Plum Creek's proposed development project for the forest looms over her plans. The latest incarnation of the construction proposal was approved by the state of Maine and preserves more wild areas than before, but legal challenges and a faltering economy still stand in Plum Creek's path. Meanwhile, Quimby's green business practices at Burt's Bees have influenced its new parent company: Clorox introduced a line of green cleaners and other environmentally friendly products that the company asserts will become a centerpiece of its future business—and a mainstay on the shelves of supermarkets and such large retailers as Wal-Mart.

Andy Frank continues to design and lobby for plug-in cars, and high gasoline prices have finally done for his vision what common sense could not accomplish on its own. Sales of sport utility vehicles have plummeted, as Americans sought out the most fuel-efficient models on the market. Suddenly every American car manufacturer was claiming it had a plug-in hybrid car in the works. Frank and his patents were suddenly very much in demand. And the newly elected Obama administration promised to make alternative vehicles and energy a national priority.

Terry Tamminen's plans to sidestep the federal government with a powerful state-based plan against global warming continued despite the roadblock thrown up by the Bush administration's refusal to grant a legally mandated waiver to California. Meanwhile, Tamminen began a new campaign to provide the public with a list of ten easy steps every household could undertake that would save consumers money while lowering their "carbon footprint" by 20 percent in twenty days (saving more than ten tons of greenhouse gases in the process). "It's like finding hundred-dollar bills and picking them up," the former pool man says, "and saving the world at the same time." His list follows, courtesy of the New America Foundation, a nonprofit, nonpartisan think tank that promotes "radical centrist politics":

1.   Adjust your thermostat by two degrees (cooler in winter, warmer in summer), to save one ton of greenhouse gas emissions a year.
2.   Switch from incandescent lightbulbs to compact fluores-

cents and save 300 pounds of greenhouse gases per bulb. Switching ten bulbs saves 1.5 tons of greenhouse gases and cuts household electric bills by seventy-five dollars a year.

3. Insulate your water heater with a simple thermal "jacket" and save 550 pounds of greenhouse gases a year.

4. Replace air-conditioner filters to save 350 pounds of greenhouse gas emissions a year.

5. Unplug "vampire" electronics that suck up electricity even when turned off—television sets, VCRs, DVD players, cable boxes, chargers—anything that is instant-on or that has a blinking light. The typical household will save half a ton of greenhouse gases just by making sure "off" is really off.

6. Wash clothes in cold water and save one ton of greenhouse gases.

7. Dry clothes on clotheslines and save nearly one and a half tons of greenhouse gases.

8. Take mass transit or telecommute once a week to save one ton of greenhouse gases.

9. Check tire inflation every week to increase fuel efficiency by 3 percent and save a quarter ton of greenhouse gases. (Most drivers have chronically underinflated tires, which make the engine work harder and burn more gas.)

10. Lose ten pounds—the average weight gain for Americans in the past ten years. Airlines use 350 million more gallons of jet fuel every year hauling those extra pounds.

11. *Bonus items Tamminen suggests, to substitute where necessary:* eat fresh food, not frozen food (fresh food consumes 90 percent less energy); eat less beef (the production of beef, pound for pound, uses up more energy than any other food); avoid bottled water and disposable grocery bags; buy local produce and other foods to avoid the 1,300 miles the average American meal travels on its way to the dinner table, using fossil fuels all the way.

Carole Allen continues to line up volunteers to find and protect the sea turtles' nests on South Padre Island. The nesting season of 2008 set a new record for hatchlings, and the numbers of sea turtles harmed by shrimping remained at historic lows. Slowly, Allen says, the Kemp's ridley sea turtle is coming back.

Ted Turner bought a solar energy company, positioned it for growth, and then sold it to a larger company controlled by the Walton family (of Wal-Mart fame), which intends to install solar panels on the roofs of its stores. He continued to lobby for a national mobilization against global warming, analogous to the extensive retooling that the nation embarked on when it entered World War II. "It's the seventh inning and we're down by two runs," Turner says. "It's a tough spot, but we can play our way out of it."

In Chile, Doug and Kris Tompkins build their parks and acquire new lands, and continue to dispense millions of dollars in grants to South American environmental groups and publications, nurturing a green movement on a continent where such a movement scarcely existed fifteen years ago. Now they are fighting hand in hand against a proposal to build massive dams on the wild rivers of Patagonia, which would flood pristine natural landscapes in and around the Tompkinses' vast holdings. Doug thinks there's a fifty-fifty chance the dams can be stopped. International organizations are joining the cause, begging Chile not to despoil its most beautiful and important landscape; Robert F. Kennedy Jr. made a personal appeal to the president of Chile to explore solar and wind power facilities before resorting to dams to generate power. America had found out the hard way that such dams were not worth the expense and environmental ruin, he argued.

In the midst of the battle, nature took matters into its own hands. An enormous volcanic eruption in May 2008 near the town of Chaiten, at the southern border of Tompkins's Pumalin Park, destroyed the town and forced evacuations from the area, although the park itself weathered the crisis with little damage. The proposed dams have been put on hold, but pressure to build a new transcontinental road through Pumalin, a project that the Tompkinses have long resisted, is mounting because of the difficulty residents of Chaiten faced while trying to evacuate. Doug Tompkins still thinks ferries are the best alternative for travel through the area—they offered a far safer escape from the volcano than the proposed road, which would have passed right by the lava flows. But he understands that emotions are running high and has been careful not to appear to be an obstacle to matters of perceived public safety. "We don't have the intention or power to stop public works," he told the Chilean press. "We are just offering our position."

In the United States, Tompkins has started a new campaign against off-road all-terrain vehicles—ATVs—which he considers to be among the most senselessly destructive technological developments in man's "disregard for nature." He continues to bankroll the Center for Biological Diversity's campaign to limit ATVs on ecologically sensitive public lands. In late 2007, his Foundation for Deep Ecology published *Thrillcraft*, a collection of essays and photographs documenting the damage to the environment caused by what Tompkins, in a blistering foreword to the book, calls "slob recreation."

A different, mellower, even shy Tompkins left Chile for a short foray to Washington, D.C., in the fall of 2007. He and the more voluble Kris were feted by the International Conservation Caucus Foundation, and by Secretary of the Treasury Hank Paulson and his wife, Wendy, both respected conservationists. The Tompkinses received the foundation's "Good Steward Award" for their conservation work in Patagonia. The event was called a gala, and it was a classic assemblage of Capitol glitterati—diplomats, politicians, lobbyists, celebrities. Tompkins looked stiff and uncomfortable in a suit and tie he clearly no longer felt comfortable wearing after years in more rustic surroundings; he listened affably as the actor Robert Duvall engaged him in a lengthy conversation about the relative merits of U.S. beef versus leaner Argentinean steaks. Tompkins tactfully didn't mention that he really doesn't care for beef of any kind, though as a resident of Chile, where barbecues are a mainstay of social life and a refusal to sample the beef-laden national cuisine would be a terrible insult, he has abandoned his vegetarian preferences when circumstances require it.

When it was time to accept the award, Tompkins humbly and hesitantly spoke about the political opposition he and Kris have had to deal with because of the "passions" that conservation arouses whenever land is converted from development to preservation. He saw himself and his wife as part of a grand tradition, he told the gathering, one that might seem like a last-ditch effort to save bits and pieces of the environment, but that is in truth part of something bigger, something that runs deep and that can, he suggested, be part of a worldwide project to save the planet.

"This is the long tradition of conservation philanthropy, which is really a true American tradition," he said. "It started over a century ago. Nearly every one of our seventy-seven national parks in this country, in part or in total, . . . was created through private philanthropy."

He paused a moment, not really seeing the crowd before him, men and women resplendent in their evening gowns and perfect suits; he was seeing the national parks he knew so well, Yosemite and Yellowstone, and the new ones he was building—places where he truly felt connected. He suddenly looked eager to finish and to leave and get back to work, as he said that he and Kris had much more to do. "We're very proud to continue in that American tradition," he concluded.

To which Kris added, "We're in a hurry."

some resources

more on the eco barons

For the Internet supplement to this book, including photos of the eco barons and their projects, maps, background information, links to their individual Web sites, and more resources, visit http://ecobarons.wordpress.com.

For more information on the author and his work, coming events, book excerpts, and more, visit www.edwardhumes.com.

general environmental information, news, and advice

Grist: Environmental News and Commentary: www.grist.org and the related blog, http://gristmill.grist.org.

Greenwash Brigade: www.publicradio.org/columns/sustainability/green wash.

The Sietch Blog: www.blog.thesietch.org.

Green Options: http://greenoptions.com.

Climate Debate Daily: Get all sides of the global warming debate at http://climatedebatedaily.com.

living like an eco baron

Terrapass: Calculate your carbon footprint and find green products at www.terrapass.com.

NativeEnergy: Learn about and purchase carbon offsets at www.native energy.com.

Service trips: Earthwatch Institute and the Sierra Club maintain lists of volunteer vacations that put you to work on conservation and public lands projects.

> *Earthwatch:* www.earthwatch.org/expedition.

> *Sierra Club:* http://tioga.sierraclub.org/TripSearch/show-search-results.do?triptype=SV.

Treehugger: This environmental Web site's How to Go Green guide offers tips on green home buying, green dishwashers, green gift buying, greening your sex life, and more at www.treehugger.com/gogreen.php.

driving like an eco baron

EPA Green Vehicle Guide: Learn about the greenest cars in America at www.epa.gov/greenvehicles.

Plug In America: www.pluginamerica.com.

Green Car Congress: News, reports, and information on sustainable transportation at www.greencarcongress.com.

The California Cars Initiative: www.calcars.org.

Drive Green: Calculate and offset the greenhouse gas emissions for your travel at www.drivegreen.com.

eating like an eco baron

Green Daily Green Eating Guide: www.greendaily.com/2008/02/07/eating-green-an-intro.

Eat Well Guide: Find, cook, and eat sustainable food at www.eatwell guide.org.

Sustainable Table: Another excellent resource for local and sustainable food is www.sustainabletable.org.

ACKNOWLEDGMENTS

Many thanks to all those who shared with me their time, expertise, and insight—particularly Doug and Kris Tompkins, Tom Butler, Kassie Siegel, Brendan Cummings, Kieran Suckling, Peter Galvin, Roxanne Quimby, Jym St. Pierre, Andy Frank, Terry Tamminen, Carole Allen, and Peter Bahouth. Telling the eco barons' story has been a rare and eye-opening privilege.

I could not imagine completing this book without the support and guidance of my editor, Emily Takoudes; and my agent and dear friend, Susan Ginsburg. And I can't imagine completing *anything* without my partner, Donna Wares, and our Gaby and Eben.

I did have to finish without the help of my constant companion, Nikolai, our borzoi, who sat by my side daily through the last five and a half books. I miss him every day.

NOTES

introduction: the plan—buy low, sell never

1. The full species name is *Araucaria araucana*. This is one of nineteen separate araucaria species, and one of the most imperiled. It is the national tree of Chile.

2. There is broad agreement within the scientific community that the planet is now experiencing its sixth great extinction event, also referred to as the Holocene extinction event. (The last such event, 65 million years ago, was the Cretaceous/Paleogene extinction, in which half of all living species, including most dinosaurs, went extinct.) The rate of observed extinctions has accelerated since the mid-twentieth century, with human causes (development, deforestation, pollution, hunting, fishing, and climate change) primarily responsible. By 2008, scientific estimates using different methodologies ranged from 27,000 extinctions a year to as many as 140,000 a year.

3. Estimate from "Eating Fossil Fuels," by Dale Allen Pfeiffer, *From the*

Wilderness Publications, 2004. According to data on oil use cited by Pfeiffer, a geologist, oil consumption in the agricultural sector breaks down into the following categories: 31 percent to manufacture inorganic fertilizer, 19 percent for field machinery, 16 percent for transportation, 13 percent for irrigation, 8 percent for raising livestock, and the remainder for pesticide production, crop drying, and other miscellaneous industrial farming practices. This figure of 400 gallons does *not* include energy costs for packaging, refrigeration, or cooking.

1. reaching the summit

1. Information here and throughout the book on Doug Tompkins is drawn from the author's interview with him; his own writings; the Esprit Company's official history and executive biographies; interviews with his wife, Kris McDivit Tompkins; interviews with his ex-wife, Susie Tompkins-Buell; interviews with his friend and colleague Peter Buckley; various speeches and published interviews he has given; *Esprit: The Making of an Image*, by Helie Robertson, Esprit de Corp., 1985; *The Conservation Land Trust: The First Ten Years*, 2002; *The Foundation for Deep Ecology: The First Ten Years*, 1999; *Flying South: A Pilot's Inner Journey*, by Barbara Cushman Rowell, Ten Speed Press, 2002; and "Fitzroy 1968," *American Alpine Journal*, by Douglas Tompkins, 1969.

2. the empire strikes out

1. Maureen Orth, "Esprit de Corp.: For Sportswear Mogul Doug Tompkins, Image and Attitude Are Everything," *San Francisco Chronicle*, March 22, 1987; article reprinted from *GQ* magazine.
2. The argument that environmentally sound and sustainable practices cost too much is age-old and seemingly sensible, but it is at root dishonest, because it overlooks what amounts to a massive subsidy that rewards pollution and inefficiency. Pollution, climate change, and resource depletion carry enormous price tags—the costs related to cancer, heart

disease, asthma, and other problems caused by emissions from cars and polluting industries alone are staggering. These emissions sicken and kill people—they are toxic to living things and to the environment—yet government and society bear almost all the costs. If the makers of the products that caused this harm were required to pay for the actual damages they cause, which are a true cost of doing business, then it would no longer be profitable to conduct business in an unsustainable way. As things stand, the incentives greatly favor environmental damage and unsustainable practices.

3. Doug Tompkins, "Looking Backward and Forward," *Foundation for Deep Ecology: The First Ten Years*. In this essay, Tompkins also tried to explain why he—and other creative and well-intentioned business-people—could remain blind to the ecological consequences of their work:

> *As far back as 1985, my hopes and expectations in life began to shift. The excitement and involvement that came with building Esprit, improving the craft of image making, marketing, organizing, and growing a complex, multinational operation, began to lose its luster. . . . I still wonder how I could have been so focused elsewhere that I was not out there with the Earth First!ers, where my heart actually longed to be. I had become fascinated by marketing and image making, and as I look back on it now, I wonder what I was really thinking about, what captivated me so. I think in a way this ability to become narrowly focused is as fundamental a root cause of the eco-social crisis as one can find. One loses sight of the greater reality, living in that microcosmic world of marketing, advertising and global distribution.*

3. lost and found

1. DINA—the Spanish acronym for National Intelligence Directorate— was Chile's secret police, with virtual unlimited powers to detain anyone

for any reason without charges or rights whenever the nation was in a state of emergency. Because the dictator Augusto Pinochet maintained Chile in a constant state of emergency, DINA operated above the law and above constitutional restraints. Human rights abuses were common and included torture, rape, and assassination. Despite this record, CIA documents made public in recent years reveal that the director of DINA was a paid "asset" of the CIA. DINA maintained a notorious facility, Colonia Dignidad, a concentration camp where political prisoners were tortured and used as guinea pigs for biological weapons experiments.

The media attacks on Tompkins and his Pumalin Park project would become so severe that one reporter asked in 1994, "What would you say to those who think Pumalin could become another Colonia Dignidad?" Tompkins said the question was absurd, but it was perhaps the most severe accusation that could be made in Chile, for no vestige of the Pinochet regime was more feared and despised.

4. image and reality

1. Region X, also called Los Lagos, is one of fifteen regions in Chile, each designated by a Roman numeral. Regions are the largest political subdivisions in the country, and they are in turn divided into provinces. Much of Patagonian Chile lies in Region X.

2. Argentina and Chile, whose border runs through Tompkins's land, have bickered since the early 1800s over ownership of three islands in the Beagle Channel at the continent's southern tip. With valuable oil and fishing rights at stake, an international court finally ruled in favor of Chile in 1977, but Argentina declared the decision void. Soon, opposing naval squadrons were facing off in the Beagle Channel, guns at the ready. War was averted only when Pope John Paul II intervened and insisted that he be allowed to mediate. After six years of acrimonious negotiations, the two countries signed a Treaty of Peace and Friendship at the Vatican. Two decades later, there are still old minefields in remote parkland near the border. When it became known that Tompkins was

buying up land in Argentina as well as Chile for conservation, and that he might be enjoying a friendlier relationship with Buenos Aires than with Santiago, the allegation that he posed a national security risk with his extensive holdings on the border did not sound so outlandish to Chileans' ears.

3. The coverage of Doug Tompkins was by no means universally positive: After granting extensive access to William Langewiesche of the *Atlantic Monthly*, who would later achieve greater prominence for his superb reporting from the World Trade Center after 9/11, Doug Tompkins was devastated by the dark, unflattering article that resulted in 1999: He was portrayed as a messianic character straight out of Conrad's *Heart of Darkness* who loved nature but would just as soon have no people around to disturb it.

5. thinking like aldo

1. In a special message to Congress on February 8, 1972, President Richard Nixon explained his rationale for the Endangered Species Act and other major environmental legislation adopted during his administration:

This is the environmental awakening. It marks a new sensitivity of the American spirit and a new maturity of American public life. It is working a revolution in values, as commitment to responsible partnership with nature replaces cavalier assumptions that we can play God with our surroundings and survive. It is leading to broad reforms in action, as individuals, corporations, government, and civic groups mobilize to conserve resources, to control pollution, to anticipate and prevent emerging environmental problems, to manage the land more wisely, and to preserve wildness.

A summary of the Endangered Species Act (ESA) is contained in "Science and the Endangered Species Act," National Research Council, 1995. Some excerpts:

The ESA defines three crucial categories: "endangered species," "threatened" species, and "critical" habitats. Endangered species and their critical habitats receive extremely strong protection; it is illegal to take any endangered species of animal (or plant in some circumstances) in the United States, its territorial waters, or the high seas. In addition to this direct prohibition, Section 7 of the act prohibits any federal action that will jeopardize the future of any endangered species, including any threat to designated critical habitat. . . .

The strength of the ESA lies with its stringent mandates constraining the actions of private parties and public agencies. Once a species is listed as threatened or endangered, it becomes entitled to shelter under the act's protective umbrella, a far-reaching array of provisions. Critical habitat must be designated "to the maximum extent prudent and determinable" and recovery plans, designed to bring the species to the point where it no longer needs the act's protections, are required if they will promote the conservation of the species.

The key to the act's power is its absolute requirement that the decision to list and protect a species must be based solely on the best scientific data available. Economic factors may not be used to "balance" the science. Thus under the ESA a tiny fish can outweigh a massive proposed dam.

2. Scientists have concluded from the fossil record that under normal conditions, and prior to the rise of human civilization, the "background extinction rate" is about one species extinction a year for every million species on earth. No one knows how many separate species of plants, animals, and microbes there are on earth now; but if there were 10 million, this would mean that every year, ten different species would become extinct under normal conditions. However, normal conditions have not existed since the rise of civilization. The evolutionary biologist E. O. Wilson of Harvard University has estimated that there are a total of 27,000 extinctions a year worldwide; Niles Eldridge, curator at the American Museum of Natural History, has put the number at 30,000

a year. Birds provide a practical example of these numbers. There are about 10,000 known bird species in the world. At the normal background rate, there should be one bird species extinction every century. But currently, there is one bird species extinction every year—100 times more than "normal," according to Peter Raven, director of the Missouri Botanical Garden. In a paper he cowrote, "Human impacts on the rates of recent, present and future bird extinctions," *Proceedings of the National Academy of Science* (July 2006), Raven estimates that the bird extinction rate will rise to 1,000 extinctions per century (or ten a year) by the end of the twenty-first century.

6. they never saw it coming

1. The Wise Use Movement is an antienvironmentalist, pro–property rights coalition of grassroots organizations and groups funded by mining, oil, and other resource-extraction industries. Militias, survivalists, and biblical literalists are significant parts of the coalition. Its purpose is to promote the use of nature and natural resources for human benefit and to oppose attempts by the environmental movement to protect nature from human exploitation—thus the term "wise use" of nature. An article of faith of the Wise Use Movement is that rural Americans suffer disproportionately from the consequences of environmental regulation.

   The movement was founded by Ron Arnold, a former technical writer specializing in aerospace. He wrote several books criticizing environmentalism, as well as the authorized biography of Ronald Reagan's first secretary of the interior, James Watt, a hero of the Wise Use Movement.

2. *Tennessee Valley Authority v. Hill*, 437 U.S. 153 (1978). The snail darter, which had no economic or recreational value, was found to be endangered by the Fish and Wildlife Service. The Supreme Court decided that this finding, along with the prevailing scientific view that a hydroelectric dam planned for the Little Tennessee River would destroy the

snail darter's habitat and probably drive it into extinction, compelled the government to halt the dam project. The Supreme Court, in a vote of six to three, rejected arguments that Congress never intended to halt large public works projects already in progress because of concerns about endangered species. This precedent has held ever since, defining the broad and often absolute power of the Endangered Species Act.

Two ironies followed the ruling: Congress passed a law specifically exempting the dam project from the act, and the project was completed, destroying the snail darters' habitat. However, the species was later found in other river habitats—the dam area turned out not to be the only place where it lived. It was removed from the endangered list in the 1980s.

3. A sampling of the Bush administration's endangered-species policies that have been challenged or shown to be unlawful:

- The administration attempted to slash the protected habitats of dozens of endangered species, including the Mexican spotted owl and the northern spotted owl, attempted to delist the desert-nesting bald eagle despite the objections of every scientist involved with this rare bird, and moved to reduce or eliminate protections for more than a dozen other species the Center for Biological Diversity had worked for years to get on the endangered list. In the process, the administration sought to redefine "endangered" in order to disqualify species listed by past presidents, making them suddenly appear to have recovered, though under the old criteria —and according to government biologists—they remained imperiled.

- Shortly after taking office, Bush derailed a plan by the Clinton administration to reintroduce endangered grizzly bears in a wilderness area in Idaho, where they had lived for thousands of years before being hunted nearly into oblivion. Instead, the administration championed a plan to allow 41,000 acres of protected wilderness in the same area to be logged.

- In 2001, Vice President Dick Cheney, with help from the political operative Karl Rove, pressured government scientists into altering their opinion that two imperiled species of salmon in Oregon's Klamath River would be harmed by diverting river water for agriculture during a drought. Farmers then got the water they wanted—and the Klamath River saw the worst fish kill in U.S. history, with more than 60,000 chinook and coho salmon lying dead on the banks of the river. A federal judge later ruled that the administration broke the law, improperly elevating economics and special interests over science. Fishing for salmon in the Klamath has been mostly forbidden since then, because of the declining numbers of the once plentiful fish.

- The administration overrode its own scientists by seeking to remove from the endangered species list gray wolves, pygmy owls, sea turtles, snowy plovers, and the Sacramento split-tail fish. This move was accompanied by yet more bids to open up protected lands to development.

- The administration used bad science and false data on the endangered Florida panther, the official state animal, to avoid impeding development in the southwestern part of the state. Among other tricks, federal officials based their calculations on the viability of the species by assuming that every living panther was a breeding adult—in other words, that there were no juvenile, aged, or sick panthers. Then they calculated the panther's range on the basis of its daytime activity, even though the panther mostly rests during the day and hunts at night. The Florida panther is one of the most seriously endangered species in America and faces imminent extinction—only about 100 of these panthers are believed to remain in the wild.

- The president opened up America's share of the Arctic Ocean to oil drilling and exploration while simultaneously weakening protections for endangered sea life in the region, including fragile populations of the right whale, the polar bear, the ribbon seal, and the yellow-billed loon.

- The president opened 52,000 acres of Los Padres National Forest in southern California to oil and gas exploration—next to the Sespe Condor Sanctuary, one of two protected habitats for the California condor.

- The administration authorized hundreds of millions of dollars' worth of oil, gas, and mining projects in national forests and other public lands while also exempting those projects from the Clean Water Act, the Safe Drinking Water Act, and the National Environmental Policy Act.

- The president insisted that military bases be exempted from most environmental laws, including the Endangered Species Act, in the interest of national security after 9/11. Congress acquiesced—despite a finding by the Government Accountability Office that environmental regulations had not hindered military readiness in the least. Exemption in hand, the administration immediately launched a series of projects potentially devastating to endangered species, among them the 132,000 acres added to Fort Irwin in the Mojave Desert to provide extra space for live-fire war games and mock tank battles. The expansion eliminates an important habitat for the endangered desert tortoise. The army decided to physically move 800 of these reptiles to another area of the desert—where the habitat is much lower in quality and where an existing population of tortoises has been ravaged by disease.

- The administration sided with the off-road vehicle industry on air pollution standards, on noise pollution standards, and in giving the vehicles continued access to environmentally sensitive areas in national forests and wilderness areas, ignoring scientific opinions inside and outside government on the damage the vehicles cause. He also sided with snowmobile manufacturers and industry groups in overruling the National Park Service's scientists and three separate studies showing that snowmobiles in Yellowstone National Park are causing extreme damage and ruining the experience of the park for most visitors.

- Bush reversed the Clinton administration's protections for 58.5 million acres of national forest in thirty-eight states that were to be kept roadless in order to preserve wildlife and habitat. He then approved large-scale logging and timber sales in old-growth forests in Oregon and California, including the landmark Sequoia National Monument, home of the giant redwoods. The action violated existing law as well as standing court orders.

- Bush repealed rules issued in the year 2000 that had made water quality, wildlife, and recreation the top priorities in managing national forests and returned the forest service to its old practice of making logging and grazing the higher priorities.

- In 2007, Bush proposed a novel means of raising money for rural schools—by selling to developers $1 billion worth of national forest and other public wilderness areas that environmentalists had spent decades trying to protect. Bush rejected the alternative funding mechanism congressional Democrats proposed, which was to close loopholes that allow many government contractors to avoid paying taxes.

4. Other major legal victories by the Center for Biological Diversity include the following:

- It won a court order directing the Bush administration to follow a long-ignored law of 1992 requiring 70 percent of annual federal vehicle purchases to consist of alternative-fuel vehicles, and to develop rules that were supposed to ensure that 30 percent of all vehicles in the nation, public and private, would run on alternative fuels by 2010. The administration refused to comply.
- It won a federal court ruling in 2003 that the Bush administration was in violation of the Endangered Species Act and the National

Environmental Policy Act for attempting to open to off-road
vehicles thousands of previously protected acres in the California
desert, where the endangered desert tortoise and other imperiled
species would have been at risk.

- It forced the administration to concede that ten species of penguins, including the emperor penguin, were endangered, then
sued again when the government missed its deadline for protecting
those species.

- It stopped activities by off-road vehicles on 50,000 acres of Algo-
dones Dunes in the California desert in 2006. This is the largest
dune ecosystem in the nation and the habitat for a broad range of
endangered desert animals and plants.

- It joined forces with an evangelical environmental group, Christians Caring for Creation, to win protections for the endangered
arroyo toad and protections for thousands of acres of its habitat,
settling a suit involving all four of California's national forests.

- In 2003 it won a suit against the National Marine Fisheries Service
that forced federal officials to apply provisions of the Endangered
Species Act to U.S.-flagged fishing vessels operating in international waters. The immediate result was the closure of the California longline fishing fleet, which once used thirty-mile filament lines
bristling with hooks to catch swordfish and tuna. The lines also
killed thousands of marine mammals, seabirds, and, most critically, leatherback sea turtles, an ancient species of 1,000-pound,
twelve-foot reptiles facing imminent extinction. The precedent on
longline fishing led the government to conduct greater environmental reviews for all high seas fisheries and to limit floating gill
nets off the California coast.

- In 2002 it won a temporary restraining order against the National Science Foundation that stopped U.S. research vessels from
conducting seismic surveys in the Gulf of California, Mexico;
these surveys were linked to whale deaths. A settlement followed,
which requires all U.S. seismic vessels to submit to the Endangered

Species Act and other environmental reviews before conducting
research.

- In 2001 and 2002 it overturned unlawful policies of the U.S. Interior Department that allowed the government to indefinitely delay adding threatened species to the official list of animals protected by the Endangered Species Act.

- It secured 400,000 acres of critical habitat for the endangered Alameda whipsnake in northern California, only to see the Bush administration accept demands by the building industry to cut the habitat by 60 percent. This led to a second lawsuit from the center, teamed with the evangelical group, Christians Caring for Creation, that further exposed political corruption within the Interior Department's endangered-species decision-making process.

- It successfully sued to establish a 36,000-square-mile protected habitat for the most endangered mammal in the world, the North Pacific right whale, which congregates every year in the Bering Sea. The center launched an international campaign to protect the animal in 2000, and the habitat was finally established in 2006. But it may be too late—the whale has been hunted down until only about 100 are left, and the species is now threatened by pollution, oil and gas exploration, and ship collisions.

- In 2006 it won protections for the white abalone and two kinds of coral: elkhorn and staghorn. They are the first invertebrates to gain endangered-species status, and the first creatures of any kind to be listed because of effects related to climate change. A follow-up lawsuit seeking protection of the coral habitats was filed and is still in the courts. The suit is another backdoor attempt to ease the threat of global warming, one of many such efforts at the center.

5. From "Report of Investigation: Julie MacDonald, Deputy Assistant Secretary, Fish, Wildlife and Parks," an undated report made in 2007 by the U.S. Department of the Interior, Office of the Inspector General. According to the report:

MacDonald imposed a policy in which scientific information supporting species protection was kept secret while information that favored delisting species was made public, a tactic that allowed the delisting of the desert-nesting bald eagle, instead of categorizing it as a separate and endangered subspecies (a designation the center eventually got restored).

When MacDonald could not get scientists to alter their findings, she would personally rewrite the scientific reports, at times reaching conclusions exactly opposite of those her own experts had reached.

In other cases, she excised the arguably impartial analyses of government biologists with no stake in the outcome of an endangered-species petition and replaced it with scientifically suspect information supplied by industry sources who were simultaneously engaged in litigation against her own department.

After exchanging e-mails and internal information with an attorney representing business interests, MacDonald rewrote a memo on threats to populations of the bull trout in the Klamath River Basin, where government scientists had recommended carving out 300 river miles of critical habitat to stave off extinction, along with a reexamination of oil and gas leases in the area that could be contributing to the trout's demise. The original memo refers to two scientific research documents that "clearly identify current and projected threats to the species, including mortality and habitat loss, fragmentation and degradation." The rewrite by MacDonald removed references to energy leases and asserted: "The identified threats are speculative, and neither document provides substantial scientific information supporting the speculation." MacDonald was then able to force a reduction in the critical habitat by 86 percent, leaving a protected area of only forty-two miles on the Klamath.

MacDonald also shared internal documents and e-mails with the Pacific Legal Foundation, which bills its program "Humanity First" as the national leader in fighting endangered-species protections.

Employees of the fish and wildlife service reported that they dreaded dealing with this angry, abusive, and sometimes foul-mouthed official; it got so bad that the manager of the California-Nevada Operations Office for Fish and Wildlife, a very senior official in the agency, started disconnecting MacDonald during conference calls when he felt she was too demeaning to his staff, and then would refuse to call her back. The manager made the extraordinary assertion during the inspector general's investigation that he would not take anything MacDonald said or did at face value.

MacDonald intervened after field researchers with the fish and wildlife service had determined that an endangered bird, the southwest willow flycatcher, had a nesting range of 2.1 miles. MacDonald insisted that the report be rewritten using a reduced range of 1.8 miles, a seemingly arbitrary alteration—except that it was not arbitrary. MacDonald's revision ensured that the bird's reported range would not cross into her home state of California.

6. Science was often turned upside down by the Bush administration; that is how it could publicly portray logging operations as saving public forests, weakened air pollution standards as providing for cleaner air, and diminished protections for endangered species as a means of staving off extinctions. Bush's Environmental Protection Agency overruled its own scientists to block states from imposing more stringent requirements for automobile exhaust and mileage than the federal government's regulations. Bush's Justice Department attempted to deny that greenhouse gases could be regulated or even defined as pollution. James Hansen, a climate scientist at NASA and one of the world's leading experts on global warming, complained that he was censored, that his congressional testimony was doctored by political employees, and that he suffered reprisals for asserting that a tipping point will be reached by 2016 and will make destructive climate change inevitable unless greenhouse gas emissions are substantially reduced before then. Hansen has harshly criticized Bush for spending most of his presidency denying the existence of a global warming crisis, and all of his presidency resisting doing anything about it.

7. The eight decisions reviewed involved the white-tailed prairie dog, Prebel's meadow jumping mouse (two decisions—delisting and critical habitat protections), arroyo toad, California red-legged frog, Canada lynx, southwestern willow flycatcher, and twelve species of the Hawaiian picture wing fly. Some of the most blatant cases of interference— such as the delisting of the Sacramento split-tail fish, which may have affected MacDonald's own property—were not reviewed, but are the subjects of center litigation.

8. Julie MacDonald's handling of scientific questions was not unique in the fish and wildlife service during the Bush administration—other political appointees in her office made the politicization of science pervasive in the agencies responsible for the Endangered Species Act. This was documented when the Public Employees for Environmental Responsibility and the Union of Concerned Scientists sent a forty-two-question survey about scientific integrity to more than 1,400 biologists, ecologists, botanists, and other working scientists in field offices of the fish and wildlife service nationwide. The agency took the extraordinary (and possibly unconstitutional) step of forbidding employees to answer the anonymous surveys, even on their own time, but nearly 30 percent responded anyway (this would be a high response rate even without such orders). The survey revealed:

- Of those whose work involved scientific findings on endangered species, nearly half had been directed, for nonscientific reasons, to avoid findings that favored protection.
- More than half of those surveyed knew of cases where commercial interests and political intervention had led inappropriately to reversals of scientific findings.
- One in five of the scientists said they had been specifically ordered to compromise their scientific integrity by excluding or altering technical information.
- An overwhelming majority—more than 70 percent—felt that their agency was headed in the wrong direction, that it was ineffective at

protecting endangered species, and that its decision makers could not be trusted to protect species and habitats.

- Half of scientists reported morale at the agency to be poor to extremely poor; only 0.5 percent rated morale excellent.

- One scientist wrote in the comments section of the survey: "I have been through the reversal of two listing decisions due to political pressure. Science was ignored—and worse, manipulated—to build a bogus set of rationales for reversal of these listing decisions."

- Another scientist lamented: "Why can't we be honest when science points in one direction but political reality results in making a decision to do otherwise?"

Beyond the science of extinction, politicization, manipulation, and censorship of science were pervasive during the Bush administration. Pressure on scientists has also been documented in scientific reports and findings involving stem-cell research, the Plan B contraceptive, childhood lead poisoning, the health effects of the 9/11 attacks, prescription drug safety, the ineffectiveness of abstinence-only sex education, the dangers to whales posed by navy sonar tests, the dangers of certain pesticides, and, perhaps most of all, global warming. Even a bipartisan expert report on voter fraud—which disproved talking points of the White House and the Republican Party about the pervasiveness of voting fraud while supporting the Democratic Party's claims about intimidation of minority voters in some areas of the country—was censored to suggest conclusions that favored the Bush administration's political views.

An excellent compendium of cases of manipulation of science by the Bush administration, *The A to Z Guide of Political Interference in Science,* was compiled by the Union of Concerned Scientists and can be found on the Internet at www.ucsusa.org/scientific_integrity/interference/a-to-z-guide-to-political.html.

7. the polar bear express

1. There are an estimated 20,000 to 25,000 polar bears in nineteen distinct populations; about 60 percent of the animals are in Canada. The population had dwindled to half that number because of overhunting, but an international treaty limiting the hunting of polar bears had then allowed the population to grow. Opponents to the Endangered Species Act have seized on this increase as evidence that more stringent protections are unnecessary. But by 2006, five of the nineteen populations were declining again, and loss of sea ice habitat is the primary culprit. (Sources: United States Geological Survey, *Findings on Polar Bears*, nine reports and executive summary released September 7, 2007; and testimony of Kassie Siegel, April 2, 2008, U.S. Senate Committee on Environment and Public Works.)

2. Siegel's testimony, particularly her unvarnished portrait of lawlessness at the White House, seemed to shock some of the senators. Until the shift from a Republican to a Democratic majority in Congress in January 2007, hearings on such topics tended to have a very different tone. More often than not, they had been convened to call global warming the "second largest hoax ever played on the American people," as the former Republican chairman of the Senate Committee on Environment and Public Works, Senator Jim Inhofe of Oklahoma, famously declared. (The largest hoax, according to Inhofe, was the idea that the Constitution establishes the separation of church and state; he also liked to compare environmentalists to Nazis and the Environmental Protection Agency to the Gestapo.)

   Inhofe, now in the minority, joined other Republicans on the committee in attacking Siegel, claiming she was twisting endangered-species law and trying to paralyze the United States with an unprecedented regulatory burden, forcing the fish and wildlife service to consult on every power plant, road, and building project that might contribute, however remotely, to global warming. Siegel responded that there would be no

paralysis or excessive regulation, because federal agencies already conduct environmental and endangered-species reviews for their projects
all the time. All that listing the polar bear would do is force those agency
reviews—and force the White House—to stop ignoring global warming
as part of the decision-making process.

3. Kempthorne was referring to the Marine Mammal Protection Act,
which in every way that matters for the polar bear is far weaker than the
Endangered Species Act, particularly because only the latter allows for
the protection of critical habitat.

4. "To make sure that the Endangered Species Act is not misused to regulate global climate change," Kempthorne said in his press conference
on May 14, 2008, his department would "issue guidance to Fish and
Wildlife Service staff that the best scientific data available today cannot
make a causal connection between harm to listed species or their habitats and greenhouse gas emissions from a specific facility, or resource
development project, or government action."

Kempthorne appears to be arguing here that because greenhouse
gas comes from many sources rather than a single source, the Endangered Species Act has no power to protect anything. The argument is
flawed; he could just as well assert that the act cannot be used to punish
one hunter who illegally kills a polar bear, because there are many other
hunters who won't be caught.

Kempthorne's position resembles the "many causes" rationale that
cigarette companies rely on when they are sued by cancer patients:
Smokers as a group may get lung cancer at horrific rates, but this doesn't
mean that an individual smoker who gets cancer can prove cigarettes,
rather than some other factors, are at fault. This argument ignores the
fact that reductions in the production of cigarette smoke—or greenhouse gases—over time and across many cases, will inarguably save the
lives of humans and polar bears. There is no legal reason why the Endangered Species Act cannot be used to achieve such a result—this is
why Kempthorne also announced that he would order a "commonsense"

rewriting of regulations implementing the act "to provide greater certainty that this listing will not set backdoor climate policy outside our normal system of political accountability."

This was a revealing admission that contradicted Kempthorne's position on the impropriety of regulating climate change through the act: If the regulations have to be rewritten to prevent regulation of greenhouse gases, that means they have for the last thirty-five years *allowed* it.

As the Supreme Court ruled in 2007 in rebuffing the Bush administration's refusal to regulate greenhouse gases under the Clean Air Act, "Agencies do not generally resolve massive problems in one fell swoop, but instead whittle away over time."

5. Even conservative, antiregulatory analysts—including those who opposed using the Endangered Species Act to regulate climate change—saw Kempthorne's position as internally inconsistent and, on its face, illegal. Kevin A. Hassett, director of economic policy studies at the American Enterprise Institute, wrote that Kempthorne's position was "legally and ethically indefensible." On May 23, 2008, in an article in the *Financial Post* headed "Bush's polar bear legal disaster," Hassett wrote: "The fact is, if they believed that inaction was the right policy, then they should have refused to list the bear as threatened. It's ludicrous to try to have it both ways. Historians will doubtless use this cynical decision as a canonical example of what was wrong with this administration."

    Hassett predicted that the Center for Biological Diversity and its colleagues would sue Kempthorne and win, and that the case would "almost surely go down in history as the turning point in the global-warming debate."

6. The center was not completely alone in its negative view of the agreement between the Tejon Ranch Company and environmental groups. In addition to local Sierra Club members, who have decried the agreement, a group of eleven prominent condor researchers and past and present members of the condor recovery team harshly criticized the

agreement's likely effect on the endangered birds as well as the secretive process that led to the agreement.

Excerpts from the group's open letter to Governor Arnold Schwarzenegger:

As former and present participants in the condor conservation program we are firmly opposed to any development proposals for condor Critical Habitat, and we know of no evidence to support claims that the recent agreement is generally endorsed by condor experts. In fact, the agreement is almost uniformly opposed by condor experts who are independent of compensation from Tejon Ranch. Proponents have misrepresented the agreement by not revealing these negative aspects to the public, a problem we try to remedy here.

If built, this development would result in substantial harm to condors, posing a significant threat to the recovery of this well known and highly revered species. That any environmental organization might agree to such consequences is alarming and raises troubling questions about how the recent agreement was reached. . . .

Condors are sensitive to many direct and indirect threats from human activities and they uniformly avoided urban and suburban areas in historical times. A major housing development in the heart of one of their most important use areas simply should not be permitted.

Incredibly, private environmental organizations with no special authority and with very limited experience with condor issues have now agreed to a deal that would allow substantial residential development of condor Critical Habitat. Sadly this deal was based on secret negotiations from which virtually all experienced condor experts were excluded. This is the worst sort of deal-making imaginable, particularly for a species that has become a public trust. . . .

Allowing Tejon Mountain Village to be built in condor Critical Habitat would represent a victory only for unnecessary trophy-home development in the wrong place. This development would be a sad

defeat for a species in which society has invested tremendous conser-
vation resources, and an even worse defeat for the future of Critical
Habitat protection for all endangered species.

These are no grounds for celebration.

The letter was signed by Dr. Noel F. R. Snyder, U.S. Fish and Wildlife Service (USFWS) biologist in charge of condor field studies, 1980–1986, and a member of the Condor Recovery Team, 1980–1985; David A. Clendenen, condor researcher and USFWS lead biologist for condors 1982–1997, and member of Condor Recovery Team 1995–2000; Janet A. Hamber, condor biologist, Santa Barbara Museum of Natural History, 1976–present; Dr. Eric V. Johnson, field condor researcher, Cal Poly San Luis Obispo, 1978–1986; Dr. Allan Mee, postdoctoral condor researcher for Zoological Society of San Diego, 2001–2006; Dr. Vicky J. Meretsky, field biologist, Condor Research Center, 1984–1986; Bruce K. Palmer, USFWS California Condor Recovery Program Coordinator, 2000–2004; Anthony Prieto, condor field biologist, 1999–present; Dr. Arthur C. Risser, Jr., Condor Recovery Team member, 1980–1985; Fred C. Sibley, USFWS biologist in charge of condor field studies, 1966–1969; and William D. Toone, Condor Recovery Team member, 1984–1992.

8. a plum in the wood basket

1. Information on Roxanne Quimby throughout this section is drawn from the author's interview with her, supplemented by biographical material from Burt's Bees and her charitable foundations, and published interviews she has granted over the years.

2. The land that was to become Maine had a stormy, war-torn history. It was claimed at varying times by the British and French monarchies; ownership and boundaries were fought over during the French and Indian War, the American Revolution, and the War of 1812. The land eventually became part of the Massachusetts Bay Colony, then was set up as a separate territory. Maine became the twenty-third state, admit-

ted to the union in 1820, though portions of its boundaries still remained in flux until 1839, when the governor of Maine declared war
on Great Britain over a border clash with Newfoundland. This was the
only time in American history that a state declared war on a foreign
nation; the dispute was resolved by treaty without any combat.

3. The Seven Sisters were International Paper, Scott Paper, Great Northern, Diamond International, Champion International, Georgia-Pacific,
and Mead Paper (later MeadWestvaco).

4. The winding down of the timber industry in the Maine Woods initially
led to opportunities for conservation of natural areas, according to an
analysis by Sara A. Clark and Peter Howell, "From Diamond International to Plum Creek: The Era of Large Landscape Conservation in the
Northern Forest," University of Maine, Margaret Chase Smith Policy
Center (2007).

5. Congress succeeded, however, in enacting a law that forbade the president to designate any additional monuments in Wyoming without the
approval of the state's congressional delegation. That restriction, which
applies only to Wyoming, is still in effect.

6. The account of Plum Creek's lobbying and the meeting with Julie Mac-
Donald is based on evidence and memos cited in a "Notice of Intent
to Sue," dated August 8, 2007, filed by the Natural Resources Council
of Maine, the Center for Biological Diversity, and other environmental
organizations; the testimony of Daryl DeJoy, executive director of the
Wildlife Alliance of Maine, before the Maine Land Use Regulatory
Commission; "Canada Lynx Protection: How not to make decisions,"
Kennebec Journal (August 1, 2007); and statements by the U.S. Fish and
Wildlife Service director, H. Dale Hall.

9. bees and trees

1. This early recycling program was canceled when it became clear that
the benefits of recycling were outweighed by the environmental harm
caused by packaging and mailing the tiny tubes.

2. Carbon offsets are an increasingly popular method of dealing with the greenhouse gas emissions that cause global warming—either as a voluntary purchase by companies and individuals, as in the case of Burt's Bees, or as part of the "compliance market," in which companies and governments purchase offsets to meet mandates for capping greenhouse gas emissions. "Cap and trade" proposals for dealing with greenhouse gas emissions, such as those California is planning to institute, rely on carbon offsets.

A carbon offset is a kind of financial instrument; one offset represents a reduction of one metric ton of carbon dioxide or its equivalent in other greenhouse gases, such as methane. The average car getting thirty miles to the gallon and being driven 12,000 miles a year produces about 3.5 tons of carbon dioxide, so it would require 3.5 offsets to balance it out and make the driver "carbon neutral"—$20 to $50, depending on the project the offset is used to finance. Burt's Bees purchases about $25,000 worth of offsets annually.

10. your land is my land

1. All quoted testimony is from the public hearing transcript for Zoning Petition 707, Maine Land Use Regulation Commission.

11. andy frank and the power of the plug

1. Quotes and biographical information on Andy Frank are based on the author's interviews, as well as Frank's numerous published papers, articles, and speeches, and his public appearances.
2. According to the Internal Revenue Service, these tax credits were available for the following 2008 hybrid models:

Chevrolet Malibu hybrid: $1,300
Chevrolet Tahoe Hybrid SUV: $2,200
Ford Escape Hybrid: $3,000

GMC Yukon Hybrid: $2,200

Honda Civic Hybrid: $2,100

Lexus RX 400h: $1,100

Lexus LS 600h: $900

Nissan Altima Hybrid: $2,350

Saturn Aura Green Line: $1,300

Saturn Vue Hybrid: $1,550

Toyota Camry Hybrid: $1,300 if purchased before April 1, 2008;
    $650 if purchased before October 1; no rebate after October 1

Toyota Prius: $1,575 before April 1; $787 before October 1; no
    rebate after October 1

The rebate was designed to phase out for the most popular models of hybrids, which also happen to be the most fuel-efficient models—creating incentives to buy less fuel-efficient cars.

3. The Ford Model T, which at one point had 50 percent of the vehicle market in North America, had the added disadvantage of a gravity-fed fuel system that worked poorly on hills. Common practice for Model T owners was to stop, turn around, and drive backward up hills to avoid stalling. The Baker Electric had no issues with hills.

4. "Edison Batteries for New Ford Cars: Automobile Man Discusses Possibilities of New Storage Systems with Its Inventor," *The New York Times* (January 11, 1914). According to the article, Henry Ford also said, "Mr. Edison and I have been working for some years on an electric automobile which would be cheap and practicable. Cars have been built for experimental purposes, and we are satisfied now that the way is clear to success."

5. "Edison's Latest Marvel—The Electric Country House: Any One May Now Have an Electric Plant in His Own Cellar at a Comparatively Small Cost Which Will Light and Heat It and Make Housework Easy," *The New York Times* (September 15, 1912).

6. J. R. Anderson Jr., "Automobiles Complete Long Endurance Run: Trip of 1,000 Miles Including Mount Washington Climb Proves That

Edison Battery Is No Longer a Myth," *The New York Times* (October 16, 1910). The article reported dramatically on the car's performance and its power source: "It seems incredible that the power of streams and coal can be changed to an invisible force capable of being stored in the little steel cans of this battery to be drawn on at will."

7. Although it was never used extensively in cars, Edison and his heirs continued to manufacture his breakthrough nickel-iron battery quite profitably and practically unchanged for sixty years. Today it is largely forgotten, but it proved to be the most durable battery ever designed—a testament to the inventor's ability to devise technology far ahead of his time. Except for poor performance in cold weather, the Edison batteries remain superior to those used in modern gasoline cars; they were favored by industries needing long-lasting power sources in inaccessible locations—remote railroad signals and switches, mining operations, construction equipment, forklifts. Edison batteries more than fifty years old have been found in such applications, still working at nearly their original efficiency.

The late-night television host Jay Leno, an avid car collector, has a 1909 Baker Electric vehicle that still runs with its original Edison batteries.

In the 1970s, the company was finally bought by Exide, which stopped manufacturing the Edison batteries a short time later—not because anything was wrong with them, but because they lasted so long that there were no repeat sales. They were deemed to be too good to be good for business. A Chinese manufacturer still makes the same battery, using the same original (1910) internal design, but the batteries have not been used in vehicles since Edison's day, and they were largely forgotten by researchers who have sought to revive the electric car from time to time over the intervening decades.

8. A shady consortium, the Electric Vehicle Company, made matters worse, attempting to monopolize the electric taxi business in American cities through what was then called the Lead Cab Trust, named after the lead that was an essential ingredient in the car batteries. The

company sold more than 2,000 cabs, but the business sank because of poor quality control at the factory and antitrust problems in court. The company then acquired and attempted to enforce the notorious Selden patent of 1895—a patented description of a liquid hydrocarbon–fueled vehicle. On the basis of this patent, the company sought to collect royalties from all car manufacturers. Some carmakers complied, but Henry Ford sued and won a hugely publicized judgment in 1911. This outcome made him a populist hero at the time and cast the electric car consortium—which by then was bankrupt—as an enemy of progress and of the common man.

9. Gladwin Hill, "Transit Line Plot Laid to Nine Firms: U.S. Indictment Says Rule of Network in 16 States Sought—GM, Firestone Named," *Los Angeles Times* (April 11, 1947); "Transit Lines Found Guilty in Trust Case," *Los Angeles Times* (March 13, 1949); "Transit Guilt in Bus Line Case Upheld," *Los Angeles Times* (June 4, 1951).

10. In recent years, the idea of an automotive conspiracy to kill electric streetcars has been derided as paranoia, notwithstanding the conviction in 1949, later upheld by the U.S. Supreme Court. Instead, it is argued that trolleys, many of which ran on surface streets where they were slowed by snarled traffic, had become money losers in the age of the automobile—slow, cumbersome, and unable to compete. Even if that were true—and in most locations, it was not—this was a situation created by the presence of automobiles, and by civic leaders' failure to design traffic flows around existing trolley lines, rather than pitting cars against them. A company that wanted the trolley lines it purchased to stay in business would have insisted on such traffic engineering, and companies with the clout of GM, Standard Oil, and the others would have received such consideration if they demanded it. But they did not.

Another important factor was the suburban housing boom after World War II —and the failure of trolley lines controlled by auto companies to extend service to the new communities. Earlier in the century, the electric rail lines had always grown along with development, but after the war, new highways were built instead. (These highways could have incorporated light rail but usually did not, as the rail companies never sought it.) Suddenly, but not coincidentally, the nation's new suburbanites found they needed to be two-car households in order to avoid having most of the family stranded in the hinterland during the workday.

11. From *Who Killed the Electric Car?* (2006), a documentary film written and directed by Chris Paine.

12. Ovonics Battery Company, founded by the inventor Stanford Ovshinsky, finally succeeded in building on and exceeding Edison's old nickel battery, devising a new, better-performing large-format nickel metal hydride battery for a new generation of electric cars in the 1990s. The battery was announced with great fanfare. But the hoopla ended when rights to the battery were bought by General Motors and then by Chevron Oil. The batteries have been mostly unavailable for use in all-electric vehicles since then.

As GM and the other automakers fought the ZEV mandate, there was clear evidence that the battery technology was better than they were letting on. An upstart company in Massachusetts, Solectria, had used an old contract that predated the sale to GM to obtain the Ovonics batteries. It used those batteries to build a four-passenger car that resembled the streamlined EV1, and that had been crash-tested with impressive results: its light but very strong construction seemed to make it one of the safest cars on the road.

Most remarkably, the battery technology that GM claimed was inadequate for public consumption allowed the Solectria sedan to travel 375 miles on a single charge on a test track—a revolutionary accomplishment that should have validated electric vehicles once and for all.

The car had plenty of room, a full-size trunk, air-conditioning, and stereo—everything consumers demanded. In 1997, hoping to generate publicity and interest investors, Solectria followed up with a real-world drive between Boston and New York in stop-and-go traffic, a 217-mile journey that included a wrong turn, again completed on a single charge—something the big automakers adamantly claimed could not be done. As recently as April 2008, the head of research and development for GM publicly asserted that the "best battery known to mankind" could power a car only a small fraction of that distance—about forty miles—before running out of juice. That has been the public position of the major carmakers since Edison's day, and no amount of proof to the contrary seems to change it.

The Solectria project died when California killed its electric car mandate, along with the emerging market the small company had been counting on to introduce its revolutionary car.

13. The story of the California Air Resources Board's ZEV mandate is one of failed resolve. The board bowed to industry pressure and extended deadlines for increased electric car production in 1998 and again in 2001, as the car manufacturers pressed a lawsuit against the mandate. By the time the deadlines were relaxed, GM had produced 660 EV1s, Toyota had made a similar number of Toyota RAV4-EVs, and other manufacturers had contributed smaller numbers of vehicles to the ZEV fleet. Other states had expressed interest in adopting the same mandate, but the companies stopped making the cars and were letting the existing leases run out, even though California had only delayed, not abandoned, its mandate for tens of thousands of zero emissions cars.

Many of the consumers who had leased the existing electric cars came to the board to sing their praises—the speed, power, and zero emissions made them the perfect choice for California, they argued. "It's the best car I've ever owned," said one owner of a Toyota electric, a retired courtroom reporter, Linda Nicholes. She was so enthusiastic that she became president of a nonprofit electrical vehicle advocacy group, Plug-In America, which grew out of the advocacy and lobbying efforts to save the ZEV mandate; the group's executive director, Chelsea Sexton, had been part of the EV1 sales force until GM fired them all.

A showdown came on April 24, 2003. Although the vast majority of witnesses, many of them drivers of electric cars, urged the board to keep its original zero emissions mandate, the air resources board voted to do as the carmakers requested, gutting the electric car program and shifting its emphasis to the carmakers' preferred new alternative fuel, hydrogen. Instead of building hundreds of thousands of nonpolluting electric vehicles immediately, the carmakers would be allowed to fulfill their near-term obligations to cleaner air by building as few as 250 prototype hydrogen fuel cell cars, along with some modest hybrids that needed to travel only ten all-electric miles to pass muster—a ludicrous standard that Andy Frank had been beating since the 1970s. Then in eleven years, the companies each would have to put 25,000 ZEV hydrogen vehicles on the road. The carmakers happily promised to do this. The meeting was long and contentious, particularly as the board chairman gave unlimited time to the car companies' representatives to testify, but mercilessly cut off advocates of the electric car after three minutes each. Frank advocated in vain for plug-in hybrids, as he had been doing throughout the long delays and complaints from automakers.

The board majority did not question why two industries that had steadfastly resisted introducing new technologies to improve cars, mileage, and emissions would suddenly embrace a radical alternative—hydrogen—skating over the fact that the few hydrogen prototypes then in existence were fabulously expensive and could not come close to matching batteries in performance, cost, or range.

Alan Lloyd, who was then the chairman of the air resources board, led the charge to scrap the electric car program and won the majority of votes he needed. It turned out he had also accepted another position four months earlier: He was named chair of the California Fuel Cell Partnership, a public-private initiative to promote hydrogen cars that was dominated by the major automakers.

12. can a malibu pool cleaner save the world?

1. From the author's interview with Terry Tamminen.
2. Arnold Schwarzenegger's gubernatorial campaign of 2003 included an "Action Plan for California's Environment," written by Tamminen.

The plan began with this summary:

California's economic future depends significantly on the quality of our environment. We face serious environmental challenges, which have a profound impact on public health and the economy. "Jobs vs. the environment" is a false choice. Overwhelming evidence demonstrates that clean air and water result in a more productive workforce, and a healthier economy, which will contribute to a balanced state budget. Moreover, it is children who suffer disproportionate impacts of environmental toxins. Studies show that children who live near freeways, for example, suffer significantly higher asthma rates and learning disabilities. This administration will protect and restore California's air, water and landscapes so that all the people of California can enjoy the natural beauty that is California.

Specific campaign promises included an initiative to cut air pollution by 50 percent; to invest in hydrogen fuel cells and a "hydrogen highway" for a new generation of cars; to purchase clean vehicles for the state's large fleet of government cars and trucks; to launch a green building initiative statewide; and to promote solar and renewable energy at the utility level and through an extensive rooftop solar initiative, with the goal of producing 20 percent of the state's electricity through clean renewables by 2010, and 33 percent by 2020.

After the election, all the promised programs were set in motion. The goals regarding air pollution and greenhouse gases have become more ambitious, though the renewable energy targets for 2010 appear to have been unrealistic. The hydrogen highway has stalled, owing to the inability to develop practical, affordable hydrogen cars.

3. The CAFE standards regulate the fuel economy—miles per gallon—for the entire fleet of cars produced by each manufacturer. The regulation sets minimum standards, which, under President Carter, constituted an aggressive target for improving fuel efficiency. As a result, the fuel

efficiency of American cars rose from an average of 18 miles per gallon
in 1977 to 27 miles per gallon in 1984—a 50 percent increase. Standards
were relaxed by Ronald Reagan, and fuel economy remained unchanged
through 2007. A major flaw in the law was that it exempted sport utility
vehicles (SUVs) from the standards; that is why American carmakers
marketed them so aggressively. In 2007, in a case brought by the Center
for Biological Diversity, the U.S. Ninth Circuit Court of Appeals threw
out the exemption as "arbitrary and capricious."

In 2002, a study by the National Academy of Sciences found that
Americans would have consumed 14 percent more gasoline without the
CAFE standards.

4. The Kyoto Protocol is part of the United Nations' framework conven-
    tion on Climate Change, established as a worldwide effort to reduce the
    greenhouse gas emissions that cause global warming. The protocol has
    been ratified by 182 nations and all developed nations, as well as the Eu-
    ropean Union as combined entity—with the United States as a lone hold-
    out. Each nation has varying targets for global warming reductions, but
    the international average for the world's 36 top developed nations is to
    reduce greenhouse gas emissions by 5 percent below 1990 levels (which
    would be about a 15 percent reduction from 2008 levels) by the year 2012.
    After 2012, Kyoto expires and is supposed to be superseded by a new
    international agreement. Preliminary plans call for a global cap and trade
    system in which all countries, developed and undeveloped, must partici-
    pate, with the goal of cutting greenhouse gases in half by the year 2050.

5. The idea of local, regional, and state efforts to battle climate change
    has caught on as a result of federal inaction. As Tamminen worked
    on a state program, a parallel effort was getting under way involving
    hundreds of American cities that had been persuaded to combat cli-
    mate change by means of green building codes, clean energy, recycling
    programs, and converting municipal vehicles to low-emission vehicles.
    This locally based program reached a critical mass in 2005, when the
    actor and environmentalist Robert Redford convened the first of several
    "Sundance summits" for leading American mayors at his eco-resort and

conference center in Utah. Redford had argued persuasively to mayors across America that although policymakers in Washington might be paralyzed, the mayors had the power to identify and eliminate sources of greenhouse gases within their jurisdictions.

13. the turtle lady

1. The seven species of sea turtles are the flatback, green turtle, hawksbill, leatherback, loggerhead, olive ridley, and Kemp's ridley. It was once thought that the black turtle was an eighth species, but DNA analysis shows that it is a variant of the green turtle. The largest of the sea turtles is the leatherback, the only sea turtle that has no shell; instead, its protection consists of bony plates covered with skin. The leatherback, hawksbill and Kemp's ridley turtles are listed as critically endangered. All sea turtles can sense the earth's magnetic poles and use them to navigate; this ability enables them to return to their original nesting place. They are among the most ancient living reptile species.

2. National Research Council, "Decline of the sea turtles: causes and prevention" (1990).

3. This practice, known as bottom trawling or benthic trawling, is so destructive that in 2006 the National Oceanic and Atmospheric Administration banned it in more than 150,000 marine waters off the West Coast of the United States in order to protect important fisheries and habitats. The practice was curtailed in other U.S. waters as well, but the restrictions did not stop bottom trawling for shrimp in the Gulf of Mexico.

14. wild man

1. Source: IRS Form 990 for the years 2000 and 2007, Turner Foundation.

2. Source: United Nations Food and Agriculture Organization, "Livestock's Long Shadow—Environmental Issues and Options" (November 2006). According to the analysis, raising cattle—including land use practices—contributes more to global warming than transportation

does. Livestock use 30 percent of the earth's entire land surface, mostly in the form of pasture, but the figure also includes 33 percent of the world's arable land, used to produce livestock feed. The creation of pastures is a major reason for deforestation; of the lost forestland in the Amazon, 70 percent has been turned over to grazing.

Livestock accounts for 9 percent of the word's human-caused carbon dioxide (CO_2), but other greenhouse gases are far worse: livestock produces 37 percent of human-induced methane (which pound for pound causes twenty-three times as much global warming as CO_2) and 65 percent of human-related nitrous oxide (a manure-related gas that causes 296 times as much global warming as CO_2).

3. According to the United States Department of Agriculture, on the basis of 2008 estimates, the United States consumed more beef than any other country, 12.8 million metric tons. China, with its much larger population, was second, with 7.7 million metric tons. While the United States as a whole is the undisputed beef consumption leader, it is second in the world in per capita beef consumption, behind Argentina, according to the USDA. Argentineans consume about 65 kilograms per person, while U.S. citizens consume 43.8 kilograms (these figures are based on 2006 statistics).

4. Gidon Eshel and Pamela Martin, "Diet, Energy and Global Warming," submitted to *Earth Interactions* (May 2005).

epilogue: schemers and dreamers

1. Buckminster Fuller, *Operating Manual for Spaceship Earth*, Carbondale: Southern Illinois University Press (1969). Fuller, best-known for inventing the geodesic dome, was an early thinker and writer on the exhaustion of resources and the importance of building sustainability into human society and systems.

2. James Hansen, Makiko Sato, Pushker Kharecha, David Beerling, Valerie Masson-Delmotte, Mark Pagani, Maureen Raymo, Dana L. Royer, and James C. Zachos, "Target Atmospheric CO_2: Where Should Hu-

manity Aim?" (2008). Draft paper; the authors are climate scientists.

3. See World Meteorological Organization, "Climate Information for Adaptation and Development Needs" (2007); P. J. Webster, G. J. Holland, J. A. Curry, and H. R. Chang, *Science*, "Changes in Tropical Cyclone Number, Duration, and Intensity in a Warming Environment" (September 2005); Intergovernmental Panel on Climate Change, Summary of Findings (February 2, 2007); Environmental Defense Fund, "Global Warming's Increasingly Visible Impacts" (2006); and Steven W. Running, "Is Global Warming Causing More, Larger Wildfires?" *Science* (August 18, 2006).

4. Robèrt and his group of fifty scientists reached a consensus on how to pursue a national program of sustainability after twenty-one drafts of their research paper, which addressed the scientific and social issues in detail. This consensus was then distilled into four principles for building a sustainable society, "The Natural Step":

1. Nature cannot be subjected to increasing concentrations of substances extracted from the earth's crust. Example: The unfettered burning of fossil fuels—where nature had sequestered immense amounts of carbon for many millions of years—has loaded the atmosphere with toxic chemicals and greenhouse gases that now threaten life on earth with starvation, disease, and extinction.

2. Nature cannot be subjected to increasing concentrations of man-made substances that natural systems will not purge. Example: The emissions of pesticides, PCBs, mercury, chlorofluorocarbons, mine wastes, and innumerable other toxics and cancer-causing substances have contaminated food supplies, drinking water, ecosystems, and *every* human body on the planet.

3. Nature cannot be subjected to overharvesting, overfishing, and other forms of depletion beyond its ability to replenish itself. Examples: The collapse of wild salmon fisheries,

deforestation, the loss of grasslands, human-caused water shortages, and the introduction of unsustainable cattle herds to deserts and plains are just a small part of the iceberg of ecosystem overload, with ripple effects that damage nature's ability to recycle nutrients in the soil, that impair pollination of vital wild plants and crops, and that have driven mass extinctions worldwide.

4. Resources must be used fairly and efficiently in order to meet human needs globally—humans everywhere must have sufficient water, food, and shelter, or their need to survive will compel them to violate the previous three principles. Example: Logging and burning of the South American rain forest, the single largest absorber of greenhouse gases on the planet, is being driven by poverty, desperation, and greed.

The four principles may seem a matter of common sense, but modern civilization has—in the twenty-first century more than ever—consistently ignored them and, for the most part, built a human enterprise that is based on violating all four of them, extensively and habitually. Consider America's preferred form of transportation, the car, as just one example of how far we live from these principles. Even in an age of increasingly scarce and expensive gasoline, America bases an entire economy on vehicles that waste most of the energy their engines create, and that annually spew into the atmosphere millions of tons of poison—including substances known to cause cancer in humans at the molecular level, and ruinous climate change on a planetary scale. An entire ancillary industry—armies of lawyers, lobbyists, politicians, industry trade groups, labor unions, advertising firms, film companies—exists solely to convince Americans that their cars' constant waste, pollution, exhaustion of fossil fuels, and daily self-poisoning is a good, sensible, vital, sexy idea, and that we should even subsidize it with our taxes.

5. Climate Change Performance Index 2008, Germanwatch, http://www.germanwatch.org/klima/ccpi.htm.

INDEX